EXPERIMENTAL PHYSICAL CHEMISTRY

EXPERIMENTAL PHYSICAL CHEMISTRY

V D Athawale
Retired Professor
Department of Physical Chemistry
University of Mumbai, Mumbai
Maharashtra

Parul Mathur
MSc. (Chemistry)

Publishing Globally

NEW AGE INTERNATIONAL (P) LIMITED, PUBLISHERS
LONDON • NEW DELHI • NAIROBI
Bangalore • Chennai • Cochin • Guwahati • Hyderabad • Kolkata • Lucknow • Mumbai
Visit us at www.newagepublishers.com

Published by New Age International (P) Ltd., Publishers
First Edition: 2001
Reprint: 2024

GLOBAL OFFICES

- **New Delhi** **NEW AGE INTERNATIONAL (P) LIMITED, PUBLISHERS**
  7/30 A, Daryaganj, New Delhi-110002, (INDIA), **Tel.:** (011) 23253472, 23253771
  E-mail: contactus@newagepublishers.com • Visit us at www.newagepublishers.com
- **London** **NEW AGE INTERNATIONAL (UK) LTD.**
  27 Old Gloucester Street, London, WC1N 3AX, UK
  E-mail: info@newacademicscience.co.uk • Visit us at www.newacademicscience.co.uk

BRANCHES

- **Bangalore** 37/10, 8th Cross (Near Hanuman Temple), Azad Nagar, Chamarajpet, Bangalore- 560 018
  Tel.: (080) 26756823, **E-mail: bangalore@newagepublishers.com**
- **Chennai** 26, Damodaran Street, T. Nagar, Chennai-600 017, **Tel.:** (044) 24353401
  E-mail: chennai@newagepublishers.com
- **Cochin** CC-39/1016, Carrier Station Road, Ernakulam South, Cochin-682 016
  Tel.: (0484) 4051304, **E-mail: cochin@newagepublishers.com**
- **Guwahati** Hemsen Complex, Mohd. Shah Road, Paltan Bazar, Near Starline Hotel
  Guwahati-781 008, **Tel.:** (0361) 2513881, **E-mail: guwahati@newagepublishers.com**
- **Hyderabad** 105, 1st Floor, Madhiray Kaveri Tower, 3-2-19, Azam Jahi Road, Near Kumar Theater
  Nimboliadda Kachiguda, Hyderabad-500 027, **Tel.:** (040) 24652456
  E-mail: hyderabad@newagepublishers.com
- **Kolkata** RDB Chambers (Formerly Lotus Cinema) 106A, 1st Floor, S N Banerjee Road
  Kolkata-700 014, **Tel.:** (033) 22273773, **E-mail: kolkata@newagepublishers.com**
- **Mumbai** 142C, Victor House, Ground Floor, N.M. Joshi Marg, Lower Parel, Mumbai-400 013
  Tel.: (022) 24927869, 24915415, **E-mail: mumbai@newagepublishers.com**
- **New Delhi** 22, Golden House, Daryaganj, New Delhi-110 002, **Tel.:** (011) 23262368, 23262370
  E-mail: sales@newagepublishers.com

ISBN: 978-81-224-1336-6
C-23-12-13892

Printed in India at Ajit Printing Press, Delhi.

NEW AGE INTERNATIONAL (P) LIMITED, PUBLISHERS
7/30 A, Daryaganj, New Delhi - 110002
Visit us at **www.newagepublishers.com**
(CIN: U74899DL1966PTC004618)

PREFACE

Whatever you vividly imagine, ardently desire, sincerely believe and enthusiastically act upon, must inevitably come to pass.

– Paul J. Meyer

The importance of experimental chemistry in chemical education cannot be overstated. Practicals reinforce, and not merely support, the theoretical concepts taught in the classroom. Even a brief perusal of the history of chemistry, or for that matter science in general, will show that the field was born and nurtured in the laboratory, and not in the mind of an armchair theorist. Modern day practicals teach young students how to solve a problem and arrive at a conclusion. As such, they train students in the art of tackling the real challenges that lie ahead in life. They also serve as an acid test for competence in the field of chemistry.

How this book is organised

This book is organised into thirteen sections, each dealing with a particular area in physical chemistry. Each section starts off with a short biography of a famous scientist associated with that field. The theory behind the experimental work is then covered, followed by the experimental procedures themselves. A few review questions help you to gauge your understanding of the topics covered. Each section has its own appendix that contains useful data, hints to solve the review questions and the expected experimental results. Each section is designed to be a self-sufficient unit found in one place in the book.

There are three additional appendices at the end of the book. Appendix A deals with SI quantities, constants and their units. Appendix B contains the Periodic Table of Elements. Appendix C lists the practical physical chemistry syllabus of the University of Mumbai for both years of M. Sc.

Who should read this book?

This book is aimed at students of post-graduate physical chemistry. It has been written keeping in mind experimental conditions generally found in Indian laboratories (climatic as well as budgetary). The experiments in this book are a representative selection of most physical chemistry syllabi in use in Indian universities today.

This book gives you all the background theoretical knowledge as well as the practical guidance you need to successfully carry out experiments in physical chemistry.

In this book you will find the expected experimental results, which will allow *you* to judge the success of your own experiments. Thomas Alva Edison once said, "Success is 99% perspiration and 1% inspiration." Armed with all the support you will find in this book, we encourage you to work hard to get the right results.

A word of warning is needed here. Don't look ahead to see the expected results and then manipulate your results to match them. This will defeat the whole purpose of doing the experiment. If your results are very different from those expected, try to find out what went wrong and then repeat the experiment till you get the results. Remember, there are no short cuts to success. There is an old Chinese proverb that sums up this advice very well. "I hear and I forget; I see and I remember; I do and I understand."

In this book you will also find a few review questions at the end of each chapter. Finding out the correct answers on your own will go a long way in reinforcing your learning though if you are really stuck, you can always take a look at the hints provided. Again, don't jump straight from the questions to the hints. One must learn by doing things for, though you think you know it, you cannot be certain until you try.

Tell us what you think

As the reader of this book, you are our most important critic and commentator. We value your opinion and want to know what we're doing right, what we could do better, and any other words of wisdom you're willing to pass our way.

You can send us your comments by post, email, telephone or fax. When you write, please be sure to include your name, email address and postal address. Our contact information is as follows:

Mail Dr. V. D. Athawale
Department of Chemistry
University of Mumbai
Vidyanagari
Santacruz (E)
Mumbai 400098
India

Email vilasda@yahoo.com
Phone 91-22-6527958 extension 567
Fax 91-22-6528547

CONTENTS

Josiah Willard Gibbs

Josiah Willard Gibbs was born on February 11, 1839 in New Haven Connecticut, USA. In 1863, Yale University conferred on him the first doctorate of engineering in the United States. From 1866 to 1869 he studied in Europe where he was influenced by Kirchhoff and Helmholtz.

Gibbs returned to Yale in 1869 and, two years later, was appointed professor of mathematical physics. He did not publish his first work until 1873 when he was 34 years old.

In 1876 Gibbs published the work for which he is most famous: *On the Equilibrium of Heterogeneous Substances*. This was hailed as "an event of the first importance in the history of chemistry" but many years passed before its value was generally known, mainly because "its mathematical form and rigorous deductive processes make it difficult reading for anyone, and especially so for students of experimental chemistry, whom it most concerns. ..."

Gibbs published five papers on the electromagnetic theory of light between 1882 and 1889. His *Elementary Principles in Statistical Mechanics* provided a mathematical framework for quantum theory and Maxwell's work.

Gibbs died on April 28, 1903. In 1923, the American Mathematical Society named a lecture series in his honour where an annual lecture has been given by a distinguished mathematician almost every year since inception.

1 The Phase Rule

1.1 Introduction

Substances are rarely found in pure form. Our environment is a mixture of all forms of matter: the earth's crust is a mixture of soils, rocks and minerals; seas, rivers and lakes are waterbodies that contain dissolved gases, living organisms and salts; air consists of gases, water vapour, and dust particles. Such mixtures are generally not uniform in composition. For example, the earth's crust in southern Africa is far richer in diamonds than at most other places. Mixtures that are not uniform throughout in composition are known as *inhomogeneous mixtures*. This does not necessarily mean that mixtures that are uniform in composition are pure, e.g., sugar dissolves in water to form a *homogeneous mixture* of the two pure substances.

In chemistry, a mixture that is not homogeneous is said to consist of *phases*. Each phase is itself homogeneous in nature and can be separated from the other phases. For example, carbon tetrachloride and water are immiscible, and when a mixture of these two liquids is allowed to attain equilibrium, it separates into two distinct layers, the lower one consisting of pure carbon tetrachloride. A saturated solution of a salt will also consist of two phases, viz., the solution, which is a homogeneous mixture of the salt and the solvent, and the solid phase consisting of the undissolved salt. A mixture of ice, water and water vapour in equilibrium with each other consists of three phases even though the chemical species involved is the same.

A *system* is a substance or a mixture of substances that is isolated in some way from all other substances. For example, 'the water system' means that the chemical substance, water, has been separated from all other substances so that the effects of pressure and temperature on its properties can be investigated.

1.2 The Concept of Chemical Equilibrium

When a chemical reaction occurs, the concentrations of the reactants decrease with the passage of time while the concentrations of the products increase simultaneously. The *rate of a reaction* is thus defined as the amount of change in the concentration per unit time of any species involved in the reaction.

There are many reactions that tend to go in one direction under certain conditions but are capable of going in the opposite direction if the reaction conditions are changed. In other words, the products formed by the reactants may react among themselves to give back the original reactants. When this occurs, the reaction is said to be *reversible*. Most chemical reactions, under proper conditions are reversible to some extent. Reactions in which one or more of the products separate out as insoluble or volatile species come very close to being accomplished in one direction only.

Consider a reversible reaction of the type:

$$A + B \longleftrightarrow C + D$$

Reactants A and B will react to form C and D which will simultaneously begin reacting together to give back A and B. At first, the rate at which A and B react will be much greater than the rate in the reverse direction. As more and more of C and D are formed, the rate in the reverse direction will increase. If the reactions are allowed to continue, the rate of the forward reaction will soon become equal to the rate of the backward reaction. The reaction is now said to have reached *chemical equilibrium*. This equilibrium is *dynamic* in nature because both the forward and the backward reactions are still taking place without affecting the net concentrations of the various chemicals. At this point, no further reaction will *appear* to take place, since all the chemicals are consumed as fast as they are formed.

The concept of dynamic equilibrium is not confined to chemical reactions alone. For example, when a liquid is exposed to the atmosphere, it is in dynamic *physical* equilibrium with its vapour because the rate of evaporation of the liquid is equal to the rate of condensation of the vapour. Similarly, in a saturated solution, the rate at which the solute is dissolving is equal to the rate at which it is recrystallising.

A system is said to be in *thermodynamic equilibrium* when it is in thermal, mechanical and chemical equilibrium. *Thermal equilibrium* exists when there is no flow of heat from one part of the system to another. This means that all parts of the system are at the same temperature. *Mechanical equilibrium* exists when the pressure is constant throughout all parts of the system.

1.3 The Phase Rule

The great American theoretical chemist Josiah Willard Gibbs deduced one of the most useful relationships in chemical thermodynamics called the *Phase Rule*. The Phase Rule can be used to show how various parameters like temperature, pressure, and concentration of the constituents of the system can be varied while maintaining the equilibrium between the phases. The various terms used in the study of the Phase Rule are explained below.

1.3.1 Terminology

The number of homogeneous, mechanically separable and physically distinct parts of a heterogeneous system is known as the number of **phases**, P, of the system. Each phase is separated from other phases by a physical boundary. At these boundaries, sharp changes in physical, and sometimes, chemical properties occur. The phases can be mechanically separated from each other by any method, which does not interfere with the pressure, temperature or composition of the phases, such as filtration, and separation of two liquid phases by using a separating funnel.

A **constituent** is a chemical species (an ion or a molecule) that is present in the system.

The least number of chemically distinct constituents, which can be used to express the composition of each phase quantitatively in the form of a chemical equation, is known as the number of **components**, *C*, of the system.

For example, consider the following system in equilibrium:

$$PCl_{5(s)} \ \square \ PCl_{3(s)} + Cl_{2(g)}$$

Phosphorus pentachloride decomposes to give phosphorus trichloride and chlorine gas.

There are three phases in the system: $PCl_{5(s)}$, $PCl_{3(s)}$ and $Cl_{2(g)}$. The number of components of the system is two, viz., $PCl_{3(s)}$ and $Cl_{2(g)}$. Cl_2 describes the gas phase composition completely and PCl_3 describes the solid phase composition. The third phase PCl_5 can be described in terms of the other two components by way of the chemical equation describing the reaction. Thus the number of constituents of the system is three but the number of components is only two.

Variables such as temperature pressure, and concentration of the components in the various phases are also required to be specified in order to describe the system completely. The minimum number of such independent variables that must be arbitrarily fixed in order to describe the system completely are called the **degrees of freedom**, *F*, of the system.

The Phase Rule relates the phases, components and degrees of freedom of the system by the mathematical equation

$$P + F = C + 2 \qquad (1.1)$$

1.3.2 Condensed systems and the reduced phase rule

When the equilibria between solid-liquid phases are studied in an open vessel, the effect of pressure is negligible. This is because the vapour pressures of solids are very small and hence the pressure can be considered constant. The number of degrees of freedom thus reduces by one and the Phase Rule can be modified to be

$$P + F = C + 1 \qquad (1.2)$$

This is known as the ***Reduced Phase Rule*** equation and is applicable only to systems in which solid-liquid equilibria are present. Such systems are known as *condensed systems.*

1.3.3 Phase diagrams

The number of phases of a system that can exist in equilibrium at any time depends on the conditions of temperature, pressure and composition. These conditions are determined experimentally and then summarized in pressure-temperature or temperature-composition curves. These curves are called phase diagrams of the system. Phase diagrams have very important applications in industry, especially in the fields of metallurgy, ceramics and semiconductors.

We will use the Phase Rule to study systems consisting of one, two, and three components.

1.4 One-Component Systems

In a one-component, condensed system when the number of phases is one, the Reduced Phase Rule gives

$$\begin{aligned} F &= C - P + 1 \\ &= 1 - 1 + 1 \\ &= 1 \end{aligned}$$

The only degree of freedom allowed is variation in temperature.

If the component is a solid that exists in different crystalline forms under different conditions of temperature and pressure, the component is said to exhibit *polymorphism*. In case of elements, the phenomenon is termed *allotropy*.

The sulphur system is a classic example of a one-component system displaying allotropy. Solid sulphur exists in two crystalline forms: rhombic and monoclinic. At 1 atm pressure, rhombic sulphur is stable up to 95.5°C, at which it transforms into monoclinic sulphur. The temperature at which this transformation takes place is known as the *transition temperature*. Rapid heating allows rhombic sulphur to persist in metastable equilibrium at 1 atm till its melting point of 112.8°C is reached. The melting point of monoclinic sulphur at this pressure is 119.25°C, and sulphur boils at 444.6°C.

In the phase diagram of the sulphur system shown in Figure 1.1, the curve *OC* represents the rhombic/monoclinic equilibrium, the curve *AOF* represents the rhombic/vapour equilibrium, *FCD* the rhombic/liquid equilibrium, *BC* the monoclinic/liquid equilibrium, *OB* the monoclinic/vapour equilibrium and *FBE* the liquid/vapour equilibrium. The metastable triple point *F* is observed when rapid heating does not allow sufficient time for the transition from rhombic to monoclinic sulphur to occur at its normal temperature. The broken lines represent metastable equilibrium and the full lines represent stable equilibria.

1.5 Two-Component Systems

In a two-component, condensed system consisting of only one phase, the Reduced Phase Rule gives

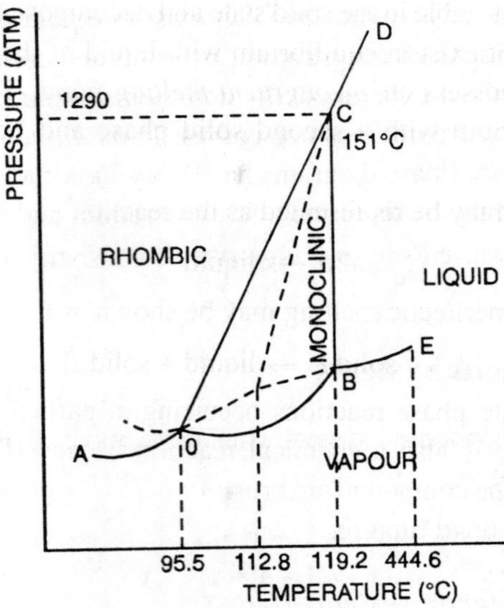

Figure 1.1: Phase Diagram of the Sulphur system.

$$\begin{aligned} F &= C - P + 1 \\ &= 2 - 1 + 1 \\ &= 2 \end{aligned}$$

The two degrees of freedom are temperature and the concentration of one of the components.

Two-component solid-liquid systems can be classified into various types depending on their solubility and on the nature of the solids separating from the solution. Table 1.1 illustrates this classification.

Table 1.1: Classification of two-component solid-liquid systems

Class	*Type*	*Description*
1. Eutectic systems	I (*a*)	Formation of simple eutectic, liquids miscible
	I (*b*)	Formation of monotectic, liquids partially miscible
	II	Formation of compound with congruent melting point
	III	Formation of compound with incongruent or meritectic melting point
2. Completely miscible solid solutions	I	Continuous series of solid solutions
	II	Solid solutions exhibiting a minimum melting point
	III	Solid solutions exhibiting a maximum melting point
3. Partially miscible solid solutions	I	Peritectic system of two solid solutions
	II	Eutectic system of two solid solutions

A compound is said to have a *congruent melting point* if the solid substance can exist in equilibrium with liquid of the same composition. Certain compounds are not stable in the solid state and decompose below their melting point. They cannot exist in equilibrium with liquid of the same composition and are said to possess an *incongruent melting point.* The solid phase can establish equilibrium with a second solid phase and liquid of a different composition.

True melting may be represented as the reaction

$$\text{solid} \rightarrow \text{liquid}$$

Peritectic or meritectic melting may be shown as the change

$$\text{solid I} \rightarrow \text{liquid} + \text{solid II}$$

Both constitute phase reactions occurring at particular temperatures. A *meritectic melting* is also a chemical reaction because the actual chemical bonds present in one compound are broken whereas *peritectic melting* involves no such chemical bond rupture.

1.5.1 Simple eutectic systems (Class 1 Type I (a))

Consider the system to be consisting of a mixture of two pure components *A* and *B* such that they are completely miscible with each other in the liquid phase but are mutually immiscible in the solid phase. A typical phase diagram of such a system is in Figure 1.2.

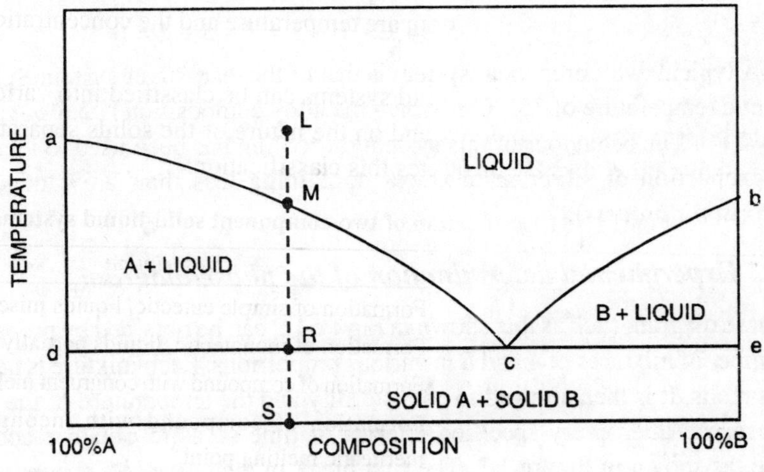

Figure 1.2: The phase diagram of a typical two-component, simple eutectic system.

- In the figure, point *a* represents the melting point of pure *A*, and point *b* represents the melting point of pure *B*.
- All the points on the curves *ac* and *bc* represent the melting points of mixtures of *A* and *B* in various proportions. These curves are called *freezing point curves* or *melting point curves*, depending upon the

context in which they are discussed. They represent the composition of the liquid phase in equilibrium with solid and hence are called *liquidus curves*.

- The curve *dce* represents the composition of the solid phase and is known as the *solidus curve*.
- The region above the curve *acb* indicates the temperatures at which the liquid mixture of *A* and *B* is stable.
- The region *acd* indicates the region of stability of the liquid mixture of *A* and *B* along with the solid *A* that has crystallised out.
- The region *bce* indicates the region of stability of the liquid mixture of *A* and *B* along with the solid *B* that has crystallised out.
- Point *c* is the point of intersection of the two freezing point curves *ac* and *bc*. It is a sharp minimum in the curve and is called the *eutectic point*. The temperature at this point is called the *eutectic temperature* and the composition of the mixture at this point is known as the *eutectic composition*. The significance of the eutectic point lies in the fact that *no mixture of A and B has a melting point lower than the eutectic temperature*.

  The eutectic point is an *invariant point* on this temperature-composition curve. This can be verified by applying the Reduced Phase Rule equation to this point. At the eutectic point, there are three phases (one liquid and two solid) and two components. Therefore, the number of degrees of freedom is

$$F = C - P + 1 = 2 - 3 + 1 = 0$$

A typical two-component system is that of the lead-silver system. It has a eutectic temperature of 305°C at which the alloy composition is 2.6% Ag and 97.4% Pb. The behaviour of this system on cooling has been used to increase the proportion of silver in mixtures containing less than 2.6% of silver (*Pattinson's process*).

1.5.2 Experimental determination of the phase diagram

A phase diagram such as that shown in Figure 1.2 can be obtained by preparing a number of mixtures of *A* and *B* in various proportions. Each mixture is heated till it melts. It is then allowed to cool steadily and the temperature of the melt is observed after every specified interval of time so as to obtain a cooling curve, as shown in Figure 1.3. The first break in the curve, *M*, occurs at the freezing point of the particular mixture. The second break, *R*, where the temperature remains constant, gives the eutectic point. The freezing points of the various mixtures are plotted against their composition to obtain the phase diagram of Figure 1.2.

On cooling a liquid mixture of *A* and *B*, solid *A* or solid *B* starts to separate out when the temperature reaches a point on the curve *ac* or *bc*, respectively

(Figure 1.2). The solid continues to deposit as the temperature falls, but when the eutectic point is reached, and both solids separate, the temperature remains constant. The separation of solid *A* or *B* is always accompanied by the evolution of heat, equal in amount but opposite in sign to the heat of fusion. Consequently, at a point equivalent to *M* in the Figure 1.2, the rate of cooling will be diminished. The cooling will thus continue from *M* to *R* at a slower rate than before until the eutectic point is attained, when the temperature remains constant.

If the temperature of the system while cooling starting with the molten liquid *L* were plotted against the time, a *cooling curve* such as that shown in Figure 1.3 would be expected. From *L* to *M* the liquid cools rapidly. At *M* solid commences to separate and the rate of cooling is slowed down. The cooling curve thus shows a break at the temperature *M*, which is the freezing point for the mixture. At *M* solid begins to form. As the composition of the liquid phase changes, the temperature falls from *M* to *R*. At *R*, the eutectic point, the temperature remains constant along *RR′* until solidification is complete. When no more liquid is left, cooling of the solid proceeds from *R′* to *S*.

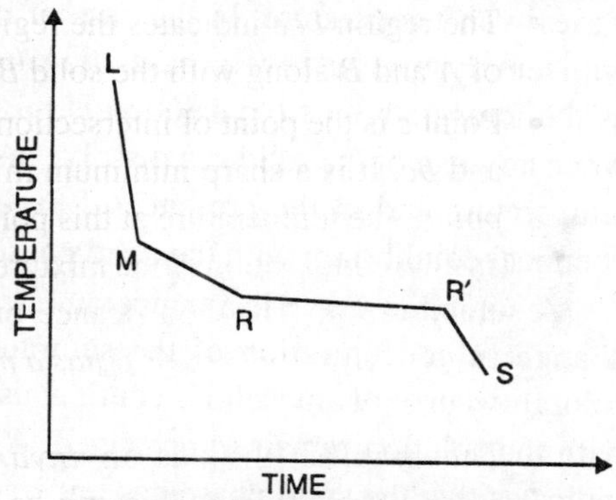

Figure 1.3: A typical cooling curve.

From the cooling curve for a mixture of any definite composition, it is thus possible to obtain the freezing point of that mixture and the eutectic temperature of the system. The freezing point varies with the composition but the eutectic point remains the same throughout (Figure 1.4). The closer the composition of the original system is to that of the eutectic, the shorter will be the portion *MR* and the longer the halt *RR′* at the eutectic temperature. If the liquid mixture coincides with the composition at the eutectic point, the region *MR* will not exist. The cooling curve will then not show a break until the eutectic temperature is reached.

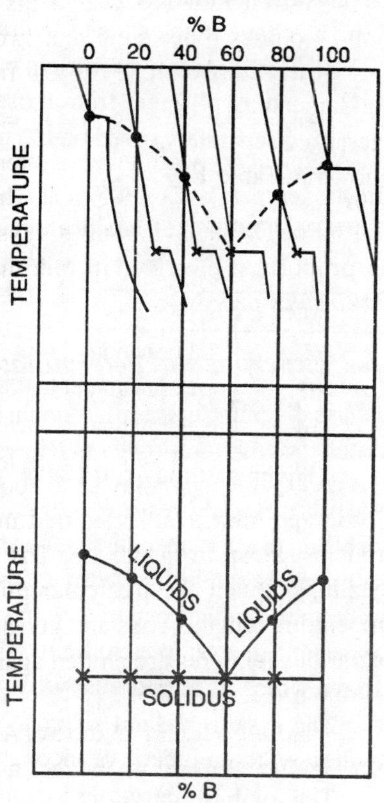

Figure 1.4: Obtaining the phase diagram.

By joining the points of first arrest (represented by circles in Figure 1.4), we obtain the liquidus curve. The solidus curve is obtained by drawing a line through the second arrest points (represented by crosses). From Figure 1.4 it is clear that this particular system forms a eutectic containing 40% *A* and 60% *B*.

1.5.3 Aqueous solutions

A solid dissolves in water to form an *aqueous solution*. Solubility generally varies with temperature. In the case of most solids, the solubility increases with temperature but several cases are known in which the solubility decreases as the temperature of the solution is raised. A *saturated solution* is formed when no more of the solid can dissolve in a given amount of the solvent at that temperature and in the presence of excess of that solid. The solution is in chemical equilibrium with the solid phase.

A solubility curve is a temperature-concentration curve representing the change in concentration of the saturated solution with temperature. The solubility curve of any solid is continuous so long as the solid phase in contact with the solution remains unchanged. A break or discontinuity in the curve indicates that the solid phase has undergone a fundamental change that may be physical in nature (e.g., an allotropic transformation) or chemical in nature (e.g., a change in the state of hydration).

Binary systems, in which one of the components is water, are not fundamentally different from those outlined in Table 1.1. However, they are classified separately based on the nature and properties of the solid phases, as shown in Table 1.2.

Table 1.2 Classification of binary systems in which one of the components is water

Type	*Description*
I	Formation of simple eutectic or cryohydrate
II	Formation of hydrate with congruent melting point
III	Formation of hydrate with meritectic or incongruent melting point

A given salt may form several hydrates depending on the conditions of temperature and pressure. At 1 atm pressure, the higher hydrate is generally stable up to a certain temperature, above which a lower hydrate or the anhydrous salt is more stable. The temperature at which the transition from one form of hydrate to another occurs is known as the *transition temperature* of the salt hydrate. For example, sodium sulphate decahydrate is stable up to 32.3°C, above which its anhydrous salt is more stable.

The system sodium-sulphate water is an example of the formation of a hydrate with meritectic or incongruent melting point.

The solubility curve of sodium sulphate decahydrate is shown in Figure 1.5. The curve *OA* shows the freezing points of various solutions up to about 5%

concentration. Curve *AE* shows the increase in solubility of the decahydrate with rise in temperature. The transition point *E* is 32.3°C at 1 atmosphere pressure. Curve *EB* represents the decrease in solubility of the anhydrous sulphate with increase in temperature.

The dashed lines *AC* and *CE′* represent equilibrium states for the heptahydrate $Na_2SO_4.7H_2O$ that is formed when the solution is cooled below 17°C without allowing sufficient time for the decahydrate to crystallise.

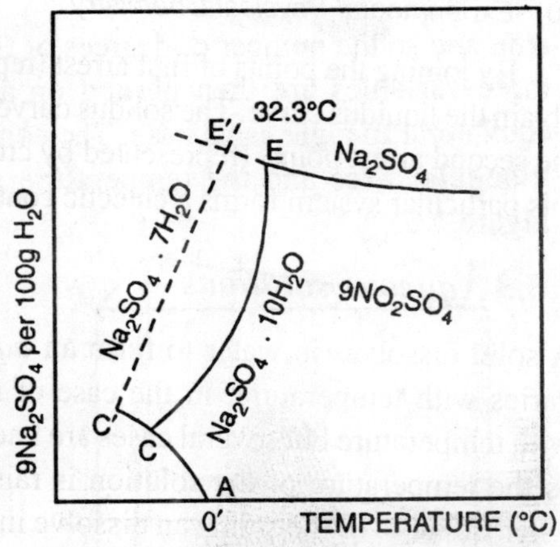

Figure 1.5: Phase diagram of the sodium sulphate decahydrate system.

1.5.4 Experimental determination of the transition temperature of hydrated salts

Phase transitions are generally accompanied by an absorption or evolution of heat. Physical properties such as colour, solubility, volume and vapour pressure are also affected. The transition temperature can be determined by measuring the change in any of these physical properties during the transition.

The solubility method depends on the fact that the two solids have the same solubility at the transition point. The method consists in determining the solubility of each form of the solid at a series of temperatures and finding the point of intersection of the two solubility curves thus obtained. Care should be taken that equilibrium is established between the solid phase and the solution by placing in a thermostat. A small amount of the saturated solution is withdrawn and weighed, and the quantity of the dissolved substance determined by analysis.

1.6 Three-Component Systems

In a three-component system consisting of only one phase, the Phase Rule gives

$$\begin{aligned} F &= C - P + 2 \\ &= 3 - 1 + 2 \\ &= 4 \end{aligned}$$

The four degrees of freedom are temperature, pressure, and the concentrations of any two out of three components. Since it is difficult to graphically represent four variables, one variable out of the four is generally considered constant. For example, the pressure is often considered fixed at 1 atm, and so the number of degrees of freedom becomes three. The remaining three variables are then plotted on a three-dimensional graph having an equilateral triangle as its base. The concentration variables are plotted on the triangular axes and the temperature is scaled vertically, as shown in the Figure 1.6.

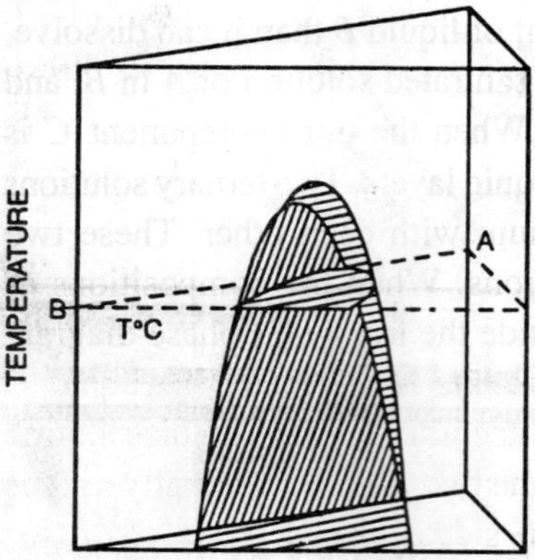

Figure 1.6: Three-dimensional representation of a three-component system having three degrees of freedom.

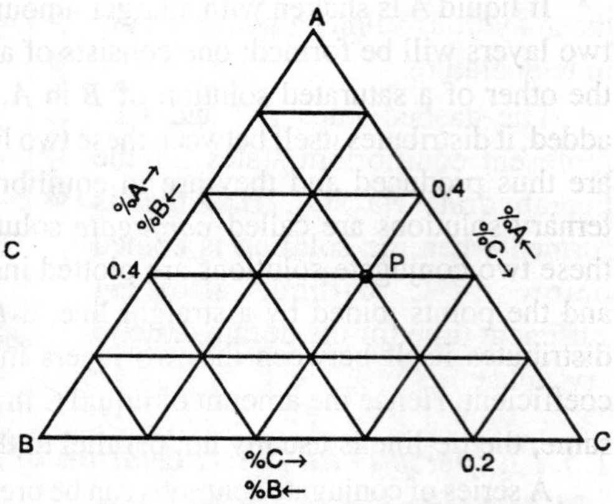

Figure 1.7: Two-dimensional representation of a three-component system having three degrees of freedom.

Any horizontal section of the right-angled prism represents a three-component system under fixed conditions of temperature and pressure, as shown in Figure 1.7.

- Each of the sides *AB*, *BC* and *AC* represents the mole fraction or percentage composition of one of the components in the mixture.
- The vertices *A*, *B* and *C* represent pure components *A*, *B* and *C*.
- Any point falling on the sides *AB*, *BC* and *BC* represents binary mixtures of *A* & *B*, *B* & *C* and *A* & *C* respectively.
- Any point falling within the triangle represents a three-component mixture.

For example, consider the point *P* in the diagram above. It represents a mixture of composition $X_A = 0.4$, $X_B = 0.2$ and $X_C = 0.4$, where X_N is the mole fraction of component *N*. It can be seen that

$$X_A + X_B + X_C = 0.4 + 0.2 + 0.4 = 1$$

which satisfies the requirement that the sum of the mole fractions of all the components must equal unity.

1.6.1 Experimental determination of the phase diagram

Consider a three-component system consisting of liquids *A*, *B* and *C*. *A* and *B* are immiscible with each other while *C* is miscible with both. A mixture of *A* and *B* at equilibrium will exist as two physically distinct phases. The addition of *C* will markedly increase the mutual solubility of *A* and *B* such that at a certain concentration of *C*, the mixture will become homogeneous.

Conversely, a mixture of A and C (or B and C) will exist as a homogeneous phase at equilibrium. The addition of the third component will markedly reduce the mutual solubility of A and C (or B and C) such that at a particular concentration of the third component, the mixture will separate out into two phases.

If liquid A is shaken with a larger amount of liquid B than it can dissolve, two layers will be formed: one consists of a saturated solution of A in B, and the other of a saturated solution of B in A. When the third component C is added, it distributes itself between these two liquid layers. Two ternary solutions are thus produced and they are in equilibrium with each other. These two ternary solutions are called *conjugate* solutions. When the compositions of these two conjugate solutions are plotted inside the triangular phase diagram and the points joined by a straight line, a *tie-line* is obtained. The liquid C distributes itself between the two layers in accordance with its distribution coefficient. Hence the amount of liquid C in the two layers is generally not the same, the tie-line is usually not parallel to the side of the triangle.

A series of conjugate solutions can be prepared by the addition of increasing amounts of liquid C to the mixture of A and B. The compositions of the successive conjugate solutions become more nearly the same and the tie-lines become increasingly shorter. A point is reached when the two solutions become identical in composition and the tie-line shrinks to a point. This point is called the *critical point* or the *plait point*. A curve joining all the points inside the triangle is called a *binodal curve*. Since the tie-lines are usually not parallel to the side of the triangle, the plait point is usually not at the summit of the curve, but somewhere below this.

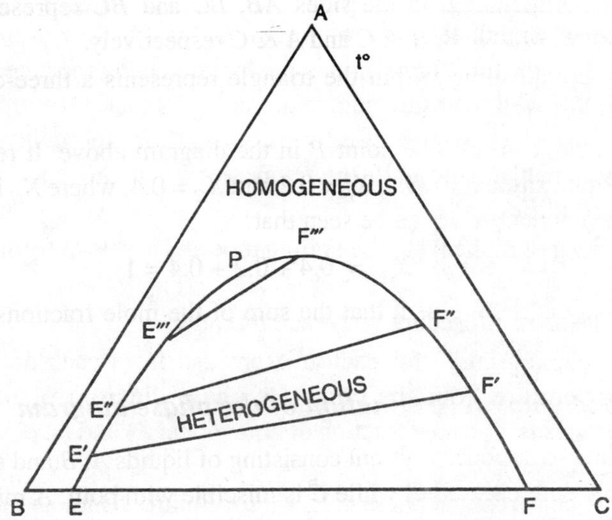

Figure 1.8: Phase Diagram of a three-component system showing tie-lines (E′F′, E″F″ and E‴F‴), plait point (P), and binodal curve (EPF).

In Figure 1.8, three tie-lines $E'F'$, $E''F''$ and $E'''F'''$ are shown.

Point *P* represents the plait point and the curve *EPF* represents the binodal curve.

A mixture represented by any point falling in the area enclosed by the binodal curve exists as two phases while a point falling outside the area of the binodal curve represents a homogeneous mixture.

EXPERIMENTS

Experiment 1.1

Aim: To determine the freezing point curve of the mixture of naphthalene and biphenyl.(or o-nitrophenol and p-toluidine).

Requirements

Cooling apparatus, boiling tubes, thermometer, naphthalene and biphenyl (or *o*-nitrophenol and *p*-toluidine).

Procedure

Prepare the following mixtures in the boiling tubes:

Set no. →	I	II	III	IV	V	VI	VII	VIII
Weight of naphthalene or or *o*-nitrophenol (g)	15	15	15	15	10	7.5	5	0
Weight of biphenyl or *p*-toluidine (g)	0	5	10	15	15	15	15	15

Take 15g of naphthalene in a boiling tube. Fit a cork with thermometer and ring stirrer to the tube. Melt the sample by immersing the tube in a beaker of hot water. Stir well to ensure complete mixing. Replace the tube in the air jacket with crushed ice to ensure uniform cooling. Stir thoroughly and record the time every half-minute until the temperature remains constant and the whole mass solidifies.

Now add 5 g of biphenyl to the boiling tube and repeat as for the previous sample.

Repeat the same procedure with each mixture.

Plot the cooling curve for each mixture and from each determine the eutectic temperature. Plot the complete phase diagram of the eutectic temperatures versus X_A (mole fraction of component *A*) and label all the areas on the diagram.

Similarly, construct the phase diagram for *o*-nitrophenol and *p*-toluidine.

Observations and calculations

Time (min)	Temperature (°C)							
	I	II	III	IV	V	VI	VII	VIII
0.0								
0.5								
...								

Set no.	% ratio of naphthalene to biphenyl (or *o*-nitrophenol to *p*-toluidine)	Eutectic temp (°C)
I	100 : 0	
II	75 : 25	
III	60 : 40	
IV	50 : 50	
V	40 : 60	
VI	33 : 67	
VII	25 : 75	
VIII	0 : 100	

Result

The nature of the curve obtained is shown in the graph. The eutectic percentage composition is ___ : ___ (naphthalene : biphenyl *or o*-nitrophenol : *p*-toluidine).

Experiment 1.2

Aim: Study of three-component system: water-acetic acid-chloroform (or water-toluene-acetic acid).

Requirements

Glass-stoppered bottles, ethanol, toluene, water.

Procedure

Prepare nine different mixtures of ethanol and toluene in the stoppered bottles with the following percentages (by weight) of ethanol: 10, 20, 30, 40, 50, 60, 70, 80 and 90. Prepare approximately 20g of each mixture by running out the calculated volumes accurately from a burette. Titrate each mixture with water, adding small amounts at a time with vigorous shaking until the first appearance of turbidity.

Calculate and tabulate the percentage by weight of each component when the solution just becomes turbid. Plot these on triangular graph paper. Draw the complete phase diagram at room temperature.

Repeat the experiment at a higher temperature.

Observations and calculations

The water-acetic acid-chloroform ternary system

1. Density of water = 1 g cm^{-3}
2. Density of chloroform = 1.478 g cm^{-3}
3. Density of acetic acid = 1.050 g cm^{-3}

Bottle no.	% of $CHCl_3$	$CHCl_3$		CH_3COOH		H_2O required for turbidity (cm^3)	
		Wt (g)	Vol (cm^3)	Wt (g)	Vol (cm^3)	At 25°C	At 35°C
1	10	2		18			
2	20	4		16			
3	30	6		14			
4	40	8		12			
5	50	10		10			
6	60	12		8			
7	70	14		6			
8	80	16		4			
9	90	18		2			

Bottle no.	% composition of final mixture by weight at 25°C			% composition of final mixture by weight at 35°C		
	$CHCl_3$	CH_3COOH	H_2O	$CHCl_3$	CH_3COOH	H_2O
1						
2						
3						
4						
5						
6						
7						
8						
9						

Calculation for % composition of final mixture by weight

Let w_1, w_2 and w_3 be the weights of the three components at turbidity.

% composition of component 1 = $100w_1 / (w_1 + w_2 + w_3)$

% composition of component 2 = $100w_2 / (w_1 + w_2 + w_3)$

% composition of component 3 = $100w_3 / (w_1 + w_2 + w_3)$

Result

As the temperature increases, the area enclosed by the binodal curve ______ (decreases/increases/remains the same).

Experiment 1.3

Aim: To determine the transition temperature of a hydrated salt.

Requirements

Solid $SrCl_2.6H_2O$ A R (or solid $MnCl_2$. $4H_2O$ A R), 12 porcelain evaporating dishes, a thermometer graduated in 0.1°C

Procedure

Prepare a saturated solution of strontium chloride hexahydrate at 100°C in 150 cm^3 distilled water taken in a beaker. The solution should always be in equilibrium with a small amount of the solid phase.

Pipette out 10 cm^3 of the saturated solution at 80°C. In order to prevent any deposition of crystals inside the pipette, the pipette should be gently heated in the Bunsen burner flame before sucking in the solution. Drain the solution in a previously weighed porcelain dish.

Cool the saturated solution with constant stirring and pipette out 10 cm^3 solution at temperatures 75, 70, 65, 60, 55, 50, 45, 40, 35°C into different, previously weighed porcelain dishes.

Allow the solution in each of the porcelain dishes to cool down and weigh them again. Evaporate to dryness the solution in each dish on a steam bath. Dry each dish in an oven at 100-105°C. Cool the dishes and weigh them again.

Calculate the weight of the solution evaporated and the anhydrous salt in each case. Calculate the solubility of the salt in g per 100 g of water. Plot the temperature versus solubility of the salt and obtain the solubility curve. The break in the curve indicates the transition temperature of the hydrated salt.

Observations and calculations

Temp °C	Wt. of dish A (g)	Wt. of dish + soln. B (g)	Wt. of dish + salt C (g)
85			
80			
...			

Temp °C	Wt. of water (B – C) (g)	Wt. of salt (C – A) (g)	Solubility (C – A)/(B – C) (g salt/100 g water)
85			
80			
...			

From the solubility curve, the break in the curve occurs at _____ °C.

Result

The transition temperature of the given hydrated salt is _____ °C.

Appendix

Review Questions

1. Is the Phase Rule applicable universally or does it suffer from some limitations?
2. With reference to Figure 1.8, when is it possible to get tie-lines parallel to the side of the equilateral triangle?

Table 1.3: Transition temperatures

Transition	*Transition temperature (°C)*
$Na_2CrO_4.10H_2O \longleftrightarrow Na_2CrO_4.6H_2O$	19.53
$Na_2CrO_4.10H_2O \longleftrightarrow Na_2CrO_4.4H_2O$	19.99
$Na_2CrO_4.6H_2O \longleftrightarrow Na_2CrO_4.4H_2O$	25.90
$Na_2CO_3.10H_2O \longleftrightarrow Na_2CO_3.7H_2O$	32.02
$Na_2SO_4.10H_2O \longleftrightarrow Na_2SO_4$	32.38
$NaBr.2H_2O \longleftrightarrow NaBr$	50.67
$MnCl_2.4H_2O \longleftrightarrow MnCl_2.2H_2O$	58.09
$SrCl_2.6H_2O \longleftrightarrow SrCl_2.2H_2O$	61.34
$CaCl_2.6H_2O \longleftrightarrow CaCl_2.4H_2O\ (\alpha)$	29.8
$CaCl_2.4H_2O\ (\alpha) \longleftrightarrow CaCl_2.2H_2O$	45.3
$CaCl_2.2H_2O \longleftrightarrow CaCl_2.H_2O$	175.3

Expected Experimental Results

Experiment 1.1

The eutectic percentage composition of naphthalene : biphenyl is 40 : 60.

The eutectic percentage composition of *o*-nitrophenol : *p*-toluidine is 50 : 50.

Experiment 1.2

As the temperature increases, the binodal curve encloses a smaller area.

Experiment 1.3

$SrCl_2.6H_2O \longleftrightarrow SrCl_2.2H_2O$	61.34°C
$MnCl_2.\ 4H_2O \longleftrightarrow MnCl_2.\ 2H_2O$	58.09°C

Hints to review questions

1. The Phase Rule is applicable to widely different physical and chemical systems. However, it does not take into account the nature or the amounts of the substances. It does not consider whether the phase reaction is physical or chemical in nature. Any change that involves the appearance or disappearance of a phase is a phase reaction. Since isotopes generally exhibit identical chemical behaviour towards changes in pressure and temperature, they are considered to be one molecular species. Hence they cannot be distinguished by the Phase Rule. Hydrogen and deuterium are exceptions because they have different physical and chemical properties.
2. When the distribution coefficient of the solute is unity (i.e., it dissolves to the same extent in both the solvents), the tie-lines will be parallel to the side of the triangle.

Further Exploration through the Internet

http://www.chem.vt.edu/chem-ed/general/thermo.html
http://van-der-waals.pc.uni-koeln.de/gpd/phaserul.html
http://jwgibbs.cchem.berkeley.edu/phase_rule.html
http://www.phasediagram.subnet.dk/

Walther Hermann Nernst

Walther Hermann Nernst was born on June 25, 1864 in former Prussia (what is now Poland). He studied mathematics and physics at the Universities of Zurich, Berlin and Graz. He did his Ph. D. from Wurzburg in 1887 on the electromotive forces produced by magnetism in heated metal plates. He thus began his career as a physicist but changed over to chemistry when he joined Wilhelm Ostwald at Leipzig University as an assistant, where van't Hoff and Arrhenius were already established.

Nernst's early studies in electrochemistry were inspired by Arrhenius' dissociation theory of ions in solution. He developed methods for measuring dielectric constants and showed that solvents of very high dielectric constants promote the ionisation of substances. He introduced the concept of solubility product to explain why solids precipitate from saturated solutions. In 1889, Nernst constructed an equation, now known as the Nernst equation, which related the voltage of a cell to the chemical properties of its constituents.

In 1906 Nernst developed his Heat Theorem which was later renamed as the Third Law of thermodynamics, and for which he was awarded the Nobel Prize in 1920. In 1918 Nernst's research in photochemistry led him to the explanation of why a mixture of hydrogen gas and chlorine gas explodes when exposed to sunlight. Nernst also carried out research in solid-state chemistry and experimentally determined the specific heats of solids at very low temperatures.

Nernst was always eager to apply the results of scientific research to industry. His Heat Theorem was applied to the industrial synthesis of ammonia by Haber's process, and to astrophysical theories. He invented an electrical piano but it was not well received by musicians. He also invented a long-lived electric glow lamp with a ceramic body that is still used in some spectrophotometers as a source of infrared radiation.

Nernst was one of the earliest pioneers in the field of physical chemistry. From 1924 to 1933 he was the Director of the Physikalisch-Chemisches Institut at Berlin. He died in Germany on November 18, 1941.

2 Nernst Distribution Law

2.1 Introduction

A solute dissolves in a solvent to give rise to a solution. The extent to which a solute may dissolve in a given solvent depends on the interaction between the molecules of the solute and the solvent. For example, a salt like sodium chloride dissolves easily in water but remains insoluble in organic solvents.

A mixture of water and carbon tetrachloride will separate into two layers, with water forming the upper layer. If sodium chloride is added to this and the mixture shaken, it will be found that all the salt has dissolved in the water. But what happens if another solute, soluble in both water and carbon tetrachloride, is added? This solute will dissolve in both the solvents, the extent of its dissolution once again decided by its interaction with the solvent. For example, iodine is about 85 times more soluble in carbon tetrachloride than in water. Due to its higher solubility, most of the iodine passes into the carbon tetrachloride layer. In other words, iodine has *distributed* itself between the two layers of immiscible solvents.

The German chemist Nernst carried out a number of experiments to study the distribution of various solutes between suitable immiscible solvents and found that whenever such a distribution occurs, the ratio of the concentrations of the solute in the two solvents is always a constant value. This general observation is now known as the *Nernst Distribution Law*. It can be stated more formally as follows:

A solid or a liquid distributes itself between two immiscible solvents in such a way that the ratio of its activities in both the solvents is constant provided the following conditions are fulfilled:

- The temperature remains constant.
- The solute is present in both the solvents in the same molecular form, i.e., the solute neither dissociates nor associates in either of the solvents.
- There is no chemical reaction between the solute and either of the solvents.
- The concentration of the solute is low.

When all these conditions are fulfilled, the ratio of the activities of the solute in the two solvents is known as its *distribution coefficient*.

Table 2.1 below lists the partition coefficients of a few solutes in different solvents.

Table 2.1: Partition coefficients

Solute	*Solvent 1*	*Solvent 2*	*Partition Coefficient*
Iodine	Water	Carbon tetrachloride	85.52
Thiophene	Liquid benzene	Solid benzene	2.35
Phenol	Amyl alcohol	Water	16.0
Iodine	Carbon disulphide	Water	410
Silver (molten)	Zinc (molten)	Lead (molten)	300

2.2 Thermodynamic Derivation of the Distribution Law

The *chemical potential* of a substance in a mixture is the contribution of that substance to the total Gibbs free energy of the mixture. Gibbs free energy is a measure of the "driving force" behind a reaction. A chemical reaction proceeds spontaneously in the direction in which there is a decrease in free energy.

Just as water spontaneously flows down from a mountain to the plains, and heat spontaneously flows from a region of higher temperature to one of lower temperature, a substance in a mixture spontaneously moves from a region of higher chemical potential to one of lower chemical potential in the mixture. This happens till its chemical potential is uniform throughout the mixture, at which point equilibrium is said to have reached.

The chemical potential of a substance in a mixture is mathematically expressed as

$$\mu = \mu^\circ + RT \ln x \qquad (2.01)$$

where μ° is the chemical potential of the substance in the pure state
R is the universal gas constant
T is the absolute temperature
X is the mole fraction of the substance in the mixture

If μ_1 and μ_2 are the chemical potentials of a solute distributed between two immiscible solvents then at equilibrium, the two are equal.

$$\mu_1 = \mu_2 \qquad (2.02)$$

Assuming that the solute dissolves in the two solvents to form ideal, dilute solutions

$$\mu_1 = \mu_1^\circ + RT \ln x_1$$

and

$$\mu_2 = \mu_2^\circ + RT \ln x_2 \qquad (2.03)$$

Equation (2.02) can thus be rewritten as

$$\mu_1^\circ + RT \ln x_1 = \mu_2^\circ + RT \ln x_2$$

$$\mu_1^\circ - \mu_2^\circ = RT \ln (x_2/x_1) \qquad (2.04)$$

Since μ_1° and μ_2° are constants, the expression $(\mu_1^\circ - \mu_2^\circ)$ can be replaced by another constant, say C.

$$C = RT\ ln\ (x_2/x_1)$$
$$C = 2.303\ RT\ log\ (x_2/x_1)$$
$$(x_2/x_1) = \text{antilog}(C/2.303RT) \quad (2.05)$$

Since all the terms on the right hand side are constant, they can be collectively replaced by another constant, say K.

$$(x_2/x_1) = K \quad (2.06)$$

Since the solutions are dilute, the ratio of the mole fractions of the solute in the two solvents is equal to the ratio of their concentrations.

$$(x_2/x_1) = (C_2/C_1) \quad (2.07)$$

Combining equations 2.06 and 2.07,

$$(C_2/C_1) = K \quad (2.08)$$

Equation 2.08 is the mathematical statement of the Nernst Distribution Law. K is known as the distribution coefficient of the solute.

2.3 Deviations from the Distribution Law

One of the conditions for the Distribution Law to hold good is that the solute must not dissociate or associate in either of the two solvents. Since all solutes do not obey these rules, their behaviour shows deviation as compared to the behaviour predicted by the Distribution Law. Accordingly, the Distribution Law is modified to take into account only the molecular species common to both the solvents.

There are three possible cases of deviation:

Case I: The solute is dissociated in one solvent but not in the other.

If the solute dissociates in solvent 2 with degree of dissociation α then its dissociation coefficient is given by

$$K = \frac{C_1}{C_2\ (1 - \alpha)} \quad (2.09)$$

where C_1 and C_2 are the total concentrations of the solute in solvents 1 and 2 respectively.

Case II: The solute is dissociated in both the solvents but with different degrees of dissociation.

If the solute dissociates in solvents 1 and 2 with degrees of dissociation α_1 and α_2 respectively then its dissociation coefficient is

$$K = \frac{C_1(1 - \alpha_1)}{C_2\ (1 - \alpha_2)} \quad (2.10)$$

where C_1 and C_2 are as defined for Case I.

Case III: The solute is associated in one solvent but not in the other.

Let the solute undergo association in solvent 2 such that n molecules of the solute polymerise as follows:

$$nS \longleftrightarrow S_n$$

If K_{eq} is the equilibrium constant of this reaction and α is the degree of association of the solute in the solvent then the distribution coefficient of the solute is given by

$$K_D = K'_D(\alpha\, K_{eq}/n)^{1/n} \tag{2.12}$$

where $K'_D = C_1/(C_2)^{1/n}$

2.4 Applications of the Distribution Law

The Distribution Law can be used to study the dissociation or association of solutes in one or both the solvents. The Distribution Law can also be used to study the complex formation between a solute and a solvent. The actual equilibrium concentrations of the various species can be determined and from these values the equilibrium constant as well as the formula of the complex can be determined.

For example, consider the reaction between Cu(II) ions and ammonia molecules to give the deep blue coloured cuprammine complex.

$$Cu^{2+} + 4NH_3 \longleftrightarrow [Cu(NH_3)_4]^{2+} \tag{2.13}$$

The co-ordination number of ammonia in this complex is 4.

To study this complex, two immiscible solvents like water and carbon tetrachloride are taken. The distribution coefficient K_D of ammonia between these two solvents is determined by titrating aliquots of the solvents with standard hydrochloric acid.

$$K_D = [NH_3]_{org}/[NH_3]_{aq} \tag{2.14}$$

To determine the co-ordination number of ammonia in the complex, the distribution of ammonia between a dilute aqueous Cu(II) solution and carbon tetrachloride is studied. The concentration of free ammonia, C', and hence the number of moles, N', in the carbon tetrachloride layer is determined by titration with standard hydrochloric acid solution. The number of moles of ammonia, N'', in the aqueous layer is thus the product of its distribution coefficient and the number of moles in the carbon tetrachloride layer.

$$N'' = K_D N' \tag{2.15}$$

Knowing the total number of moles, N_T, added to the two layers initially, the number of moles of ammonia used or complex formation, N_{comp}, can be determined.

$$N_{comp} = N_T - (N' + N'') \tag{2.16}$$

The ratio of the number of moles of ammonia used for complex formation to the number of moles of Cu(II) present gives the co-ordination number of ammonia, n, in the complex.

A similar method can be used to study the complex formation between silver ions and ammonia.

$$Ag^+ + 2NH_3 \longleftrightarrow [Ag(NH_3)_2]^+$$

As another example, consider the reaction

$$KI + I_2 \longleftrightarrow [KI_3] \tag{2.17}$$

Iodine reacts with potassium iodide in aqueous solution to form the potassium triiodide complex. Iodine is sparingly soluble in aqueous solutions and tends to volatilise from them. Both these drawbacks can be overcome if iodine is stored in the form of potassium triiodide complex. The complex easily yields iodine when the equilibrium is disturbed. This property is often used to prevent loss of iodine by volatilisation in iodimetric and iodometric titrations.

To determine the equilibrium constant of the reversible reaction in equation 2.17, the distribution coefficient of iodine, K_D, between water and an organic solvent, such as carbon tetrachloride, is determined by titrating aliquots of the two layers with standard sodium thiosulphate solution.

$$K_D = [I_2]_{org}/[I_2]_{aq} \tag{2.18}$$

Then iodine is shaken with potassium iodide solution and carbon tetrachloride, and the mixture is allowed to stand until equilibrium is established. Let the concentration of KI in the mixture initially be C.

At equilibrium, the aqueous layer contains free iodine, potassium triiodide complex and free potassium iodide.

The concentrations of iodine in the two layers are determined by titration with standard sodium thiosulphate solution. Let these concentrations be C_1 and C_2 in the aqueous and organic layers respectively. Concentration C_1 represents the iodine existing in the aqueous layer as free iodine as well as the bound complex.

$$C_1 = [I_2]_{aq,\ free} + [KI_3] \tag{2.19}$$

Thus concentration of free iodine only in the aqueous layer is given by

$$[I_2]_{aq,\ free} = C_2/K_D \tag{2.20}$$

The concentration of KI_3 is given by

$$[KI_3] = C_1 - C_2/K_D \tag{2.21}$$

The concentration of free potassium iodide is thus given by

$$[K_I]_{free} = C - [KI_3]$$

$$[K_I]_{free} = C - (C_1 - C_2/K_D) \tag{2.22}$$

The equilibrium constant of the reaction is defined by

$$K_{eq} = [KI_3]/[KI]_{free}[I_2]_{free}$$

$$K_{eq} = \frac{(C_1 - C_2 / K_D)}{C_2 / K_D(C - C_1 + C_2 / K_D)} \tag{2.23}$$

EXPERIMENTS

Experiment 2.1

Aim: To determine the equilibrium constant for the reaction $KI + I_2 \longleftrightarrow KI_3$.

Requirements

Saturated solution of I_2 in CCl_4, solid KI, standard 0.05M and 0.01M $Na_2S_2O_3$ solution, 0.1M KI solution, starch solution, separating funnels, burettes, pipettes, etc.

Procedure

[A] Determination of distribution coefficient of iodine between carbon tetrachloride and water.

Prepare the following mixtures in separating funnels:

No.	Vol. of saturated I_2 in CCl_4 (cm^3)	Vol. of CCl_4 (cm^3)	Vol. of water
1	20	0	200
2	15	5	200
3	10	10	200
4	5	15	200

Shake the mixture vigorously for 30 minutes. Allow it to stand at room temperature for 15-20 minutes so as to secure complete separation of the two liquid forms.

Withdraw a 50 cm^3 aliquot of the aqueous layer from the funnel and titrate with the standard 0.01M $Na_2S_2O_3$ solution. Starch indicator should be added just prior to the end point.

To sample out the CCl_4 layer, close the top of the 5 cm^3 pipette firmly with the finger and thrust the tip through the aqueous layer. If any aqueous layer enters the pipette, gently blow it out. Withdraw 5 cm^3 of the CCl_4 layer, wipe all liquid from the outside of the pipette before transferring it to the titration vessel. Titrate with standard 0.05M $Na_2S_2O_3$ solution after adding 2g solid KI and 10 cm^3 of distilled water. End point is dark blue to colourless.

[B] To study the triiodide equilibrium

Prepare the following mixtures in separating funnels:

No.	Vol. of saturated I_2 in CCl_4 (cm^3)	Vol. of CCl_4 (cm^3)	Vol. of KI
1	20	0	200
2	15	5	200
3	10	10	200
4	5	15	200

Shake vigorously for 30 minutes to attain equilibrium as in [A]. Remove 5 cm^3 samples of CCl_4 layer and 50 cm^3 of the aqueous layers and titrate as before with the standard 0.01M and 0.05M $Na_2S_2O_3$ solutions respectively.

Observations and calculations

Standardisation of 0.01N (approx) and 0.05N (approx) $Na_2S_2O_3$ solution

Solution in burette: 0.01N (approx) or 0.05N (approx) $Na_2S_2O_3$ solution

Solution in conical flask: 10 cm^3 of 0.01N or 0.05N $K_2Cr_2O_7$ + 5 cm^3 conc. HCl + 2g solid KI

Indicator: saturated starch solution

End point: dark blue to colourless

Burette readings (cm^3)	I	II	III	Constant burette reading
Initial				
Final				______ cm^3
Difference				

$$Na_2S_2O_3 = K_2Cr_2O_7$$
$$N_1V_1 = N_2V_2$$
$$N_1 = \_\_\_\_ N$$

Determination of D

5 cm^3 organic layer titrated versus 0.05N $Na_2S_2O_3$ V_1 (cm^3)	50 cm^3 aqueous layer titrated versus 0.01N $Na_2S_2O_3$ V_2 (cm^3)	$c_1 = 0.05\ V_1/5$ (M)	$c_2 = 0.01\ V_2/50$ (M)	$D = c_1/c_2$

Mean D = ____

Determination of Equilibrium constant

5 cm^3 organic layer titrated versus 0.01N $Na_2S_2O_3$ V_1 (cm^3)	50 cm^3 aqueous layer titrated versus 0.05N $Na_2S_2O_3$ V_2' (cm^3)	Conc. of free I_2 in organic layer = c_1' = $0.01V_1'/5$ (M)	conc. of I_2 + KI_3 in aqueous layer = c_2' = $0.05V_2'/50$ (M)	Conc. in aqueous layer of I_2 only c_1'/D (M)

Conc. of KI_3 in aq layer $c'_2 - (c'_1/D)$ (M)	Conc. of KI $c_3 = 200 \times 0.1/220$ (M)	$c_3 - \{c_2' - (c_1'/D)\}$	$K = \dfrac{c_2' - (c_1'/D)}{(c_1'/D)[c_3 - \{c_2' - (c_1'/D)\}]}$ ($mol^{-1}\ dm^3$)

Mean K = ______($mol^{-1}\ dm^3$)

Result

1. The distribution coefficient of iodine between carbon tetrachloride and water is _____ .
2. The equilibrium constant of the reaction $KI + I_2 \longleftrightarrow KI_3$ is _____ $mol^{-1}\ dm^3$.

Experiment 2.2

Aim: To determine the formula of the complex formed between silver and ammonia.

Requirements

1N NH_4OH, 0.1M $AgNO_3$ solution, $CHCl_3$, 0.1N HCl.

Procedure

[A] Determination of distribution coefficient D of ammonia between water and chloroform

Prepare $50cm^3$ of each 2M, 1M and 0.5 M NH_3 solution. To each add $50cm^3$ $CHCl_3$ solvent. Shake vigorously for 30 minutes and then allow it to settle. Titrate $20cm^3$ of organic layer against 0.1 N HCl using methyl red indicator. Thus calculate the distribution coefficient D for NH_3 in $CHCl_3$.

[B] Determination of co-ordination number n

Prepare the following mixtures in separating funnels:-

No.	Vol. of $AgNO_3$ (cm^3)	Vol. of 0.1N NH_3 (cm^3)	Vol. of $CHCl_3$ (cm^3)	Conc. of $AgNO_3$ (M)
1	25.0	50.0	50.0	0.06
2	25.0	50.0	50.0	0.08
3	25.0	50.0	50.0	0.10

Shake well in a separating funnel for about 30 minutes and allow it to stand.

Titrate $20cm^3$ of chloroform layer against 0.1N HCl. End point will be yellow to pink. Thus determine the co-ordination number 'n' of silver.

Observations and Calculations

[A] Determination of distribution coefficient D of ammonia between water and chloroform

Conc. of NH_3 c' (M)	Vol. of NH_3 taken V' (cm^3)	Total conc. of NH_3 $c = c'\ V'/100$ (M)	Vol. of $CHCl_3$ taken (cm^3)	Total no. of moles of NH_3 $NT' = c'\ V'/1000$

Titration

Solution in burette: 0.1N HCl

Solution in conical flask: 20 cm^3 of organic layer + 200 cm^3 of water

Indicator: methyl red

End point: yellow to pink

Conc. of NH_3 c' (M)	Burette reading V_1 (cm^3)	Conc. of NH_3 in organic layer $c_1 = 0.1V_1/20$ (M)	No. of moles of NH_3 in organic layer $N_1 = 50c_1/1000$	No. of moles of NH_3 in aqueous layer $N_2 = N_T' - N_1$	Distribution coefficient $D = N_2/N_1$

Mean D = ______

[B] Determination of n

Conc. of NH_3 c'' (M)	Vol. of NH_3 V'' (cm^3)	Total no. of moles of NH_3 $N_T'' = c''V''/1000$	Conc. of $AgNO_3$ C_{Ag+} (M)	Vol. of $AgNO_3$ (cm^3)	Vol. of $CHCl_3$ taken (cm^3)	No. of moles of $AgNO_3$ $N_{Ag+} = 25\, C_{Ag+}/1000$

Titration

Solution in burette: 0.1N HCl

Solution in conical flask: 20 cm^3 of organic layer + 200 cm^3 of water

Indicator: methyl red

End point: yellow to pink

Burette reading V_1' (cm^3)	Conc. of NH_3 in organic layer $c_1' = 0.1\, V_1'/20$ (M)	No. of moles of NH_3 in organic layer $N_1' = 50\, c_1'/1000$	No. of moles of NH_3 in aqueous layer $N_2' = N_1'D$	No. of moles used for complex formation $N_c = N_T' - (N_1' + N_2')$	Co-ordination number $n = N_c/N_{Ag+}$

Mean n = ______

Results

The co-ordination number of silver in the silver-ammonia complex determined using chloroform as non-aqueous solvent is ______.

Experiment 2.3

Aim: To determine the formula of the complex formed between copper (II) ion and ammonia.

Requirements

0.1M $CuSO_4$, 1M NH_3(aq), standard 0.1N HCl, $CHCl_3$, methyl red indicator, separating funnels, etc.

Procedure

[A] Determination of distribution coefficient of ammonia between chloroform and water

Prepare the following mixtures in separating funnels:

No.	Vol. of 1M aq NH_3 (cm^3)	Vol. of $CHCl_3$ (cm^3)	Conc. of NH_3 (M)
1	25	25	1.00
2	25	25	0.75
3	25	25	0.50

Shake the mixture well for about 10 minutes and then allow to separate. After apparent separation has occurred, gently agitate the bottle in a rotatory fashion to detach the layers of chloroform that usually float on the water surface. Some minute droplets of one layer persist for a considerable time (up to 30 minutes) in the other, but after 5 minutes, the error in the coefficient due to this will usually be less than 5%. Titrate 20cm^3 of organic layer against 0.1 N HCl using methyl red indicator. Thus calculate the distribution coefficient *D* for NH_3 in $CHCl_3$.

[B] Determination of co-ordination number n

Prepare the following mixtures in separating funnels:-

No.	Vol. of $CuSO_4$ (cm^3)	Vol. of 0.1N NH_3 (cm^3)	Vol. of $CHCl_3$ (cm^3)	Conc. of $CuSO_4$ (M)
1	25.0	25.0	75.0	0.06
2	25.0	25.0	75.0	0.08
3	25.0	25.0	75.0	0.10

Shake well in separating funnels for about 30 minutes and allow to stand.

Titrate 50cm^3 of chloroform layer against 0.1N HCl. End point will be yellow to pink. Thus determine the formula of the copper-ammonia complex.

Observations and calculations

[A] Determination of distribution coefficient D of ammonia between water and chloroform

Conc. of NH_3 c' (M)	Vol. of NH_3 taken V' (cm^3)	Total conc. of NH_3 $c = c' V'/100$ (M)	Vol. of $CHCl_3$ taken (cm^3)	Total no. of moles of NH_3 $N_T' = c' V'/1000$

Titration

Solution in burette: 0.1N HCl

Solution in conical flask: 20 cm^3 of organic layer + 200 cm^3 of water

Indicator: methyl red

End point: yellow to pink

Conc. of NH_3 c' (M)	Burette reading V_1 (cm^3)	Conc. of NH_3 in organic layer $c_1 = 0.1V_1/20$ (M)	No. of moles of NH_3 in organic layer $N_1 = 25c_1/1000$	No. of moles of NH_3 in aqueous layer $N_2 = N_T' - N_1$	Distribution coefficient $D = N_2/N_1$

Mean D = ______

[B] Determination of co-ordination number n

Conc. of NH_3 c''(M)	Vol. of NH_3 V'' (cm^3)	Total no. of moles of NH_3 $N_T'' = c''V''/1000$	Conc. of $CuSO_4$ C_{Cu2+} (M)	Vol. of $CuSO_4$ (cm^3)	Vol. of $CHCl_3$ taken (cm^3)	No. of moles of $CuSO_4$ $N_{Cu2+} = 25\ C_{Cu2+}/1000$

Titration

Solution in burette: 0.1N HCl

Solution in conical flask: 50 cm^3 of organic layer + 200 cm^3 of water

Indicator: methyl red

End point: yellow to pink

Burette reading V_1' (cm^3)	Conc. of NH_3 in organic layer $c_1' = 0.1 V_1'/50$ (M)	No. of moles of NH_3 in organic layer $N_1' = 75c_1'/1000$	No. of moles of NH_3 in aqueous layer $N_2' = N_1' D$	No. of moles of NH_3 used for complex formation $N_c = N_T' - (N_1' + N_2')$	Co-ordination number $n = N_c/N_{Cu2+}$

Mean n = ______

Results

The co-ordination number of copper in the copper-ammonia complex, determined using chloroform as non-aqueous solvent is ______.

Appendix

Review Questions

1. What is iodometry and iodimetry?
2. Explain how bivalent cupric ion attaches four ligands in addition to two monovalent anions?
3. Why is ammonium hydroxide solution always prepared fresh?
4. In experiments 2.2 and 2.3 why is the ratio of the concentrations of ammonia to metal ion so large?
5. Give an example of a solute-solvent system that shows deviation from the Distribution Law.

Expected Experimental Results

Experiment 2.1

The distribution coefficient of iodine between carbon tetrachloride and water is 85.5 but any value in the range of 80-90 is acceptable. A large deviation from this value generally occurs because of inadequate shaking of the mixtures in the separating funnels. The equilibrium constant of the reaction $KI + I_2 \longleftrightarrow KI_3$ is of the order of $5 - 8 \times 10^2$ mol^{-1} dm^3.

Experiment 2.2

The co-ordination number of silver in the silver-ammonia complex is 2. A smaller value generally results because of errors made in the determination of the distribution coefficient of ammonia. The ammonia solution should be freshly prepared and standardized for this experiment. The distribution coefficient of ammonia between chloroform and water is around 20 to 25.

Experiment 2.3

The co-ordination number of copper in the copper-ammonia complex is 4.

Hints to review questions

1. Iodine in aqueous solution acts as a mild oxidising agent. Strong reducing agents like sodium thiosulphate, hydrogen sulphide, suplhurous acid and stannic chloride react rapidly and completely with iodine in acidic solution. Comparatively weaker reducing agents like arsenites and antimonites react completely only in neutral or faintly acidic medium. The volumetric estimation of reducing agents by the use of standard iodine solution is called *iodimetry*.

The reaction of iodine with thiosulphate is as follows:

$$2S_2O_3^{2-} + I_2 \rightarrow S_4O_6^{2-} + 2I^-$$

This reaction can be used to indirectly estimate an oxidising agent, which reacts with potassium iodide to set free an equivalent quantity of iodine. For example, copper ions react with potassium iodide as follows:

$$2Cu^{2+} + 4I^- \rightarrow 2CuI + I_2$$

All such indirect estimations of oxidising agents involving the liberation of iodine and subsequent volumetric estimations of the liberated iodine are called *iodometry*.

2. Copper forms a co-ordination complex with ammonia as ligand. The central metal ion in a co-ordination complex has a primary co-ordination number (2 in case of copper (II) ions) and a secondary co-ordination number (4 in case of copper-ammonia complex). Further reading on co-ordination complexes is recommended from any standard book on advanced inorganic chemistry.
3. Ammonia is a very volatile gas and tends to escape from its solution easily when the solution is allowed to stand for more than a day. In experiments 2.2 and 2.3 it is crucial to know the exact concentration of ammonia before and after distribution and hence chances of its escape from its solution must be minimised.
4. In experiments 2.2 and 2.3 the reaction

   $$\text{Metal} + \text{Ammonia} \longleftrightarrow \text{Metal-Ammonia complex}$$

   should proceed completely to the right. According to the Le Chatelier Principle, this can happen when the concentrations of the reactants are high. Also, it is important that all the copper (II) present in the solution should get converted into the complex and the excess ammonia is free to distribute itself between the two solvents. Since the Cu:NH_3 ratio required for complex formation itself is 1:4, a Cu:NH_3 ratio of 1:10 is usually employed.
5. When benzoic acid distributes itself between water and benzene, it exists in the aqueous phase as dissociated ions. Thus it shows deviation of type discussed in Section 2.3, Case I.

Further Exploration through the Internet

http://www.nernst.de/
http://www.chem.missouri.edu/chem15/chapter7/text.html

Svante August Arrhenius

Arrhenius was born on February 19, 1859 in Sweden. He was a child prodigy, teaching himself to read at three and finishing school as the youngest and the brightest in his class. As an undergraduate at the University of Uppsala, he worked on the properties of electricity passing through solutions. In his doctoral thesis in 1884 he proposed that the conductivity of solutions could be explained if it were assumed that molecules could break up into electrically charged fragments called ions. His data supported his explanation but the idea proved too revolutionary for his examiners' taste. His thesis was given the lowest possible passing grade.

Arrhenius accepted a position at the Technical High School of Stockholm, from where he sent copies of his thesis to other chemists in Europe. He obtained a travel grant and worked at the University of Leipzig. The older chemists were still not convinced that his theory was right but some of the younger ones were intrigued by it and started to explain their results in terms of ions and dissociation. In the late 1890s when electrically charged subatomic particles were discovered, Arrhenius' theory suddenly made sense.

Arrhenius became professor of chemistry at the University of Stockholm in 1895. In 1903, he was awarded the Nobel Prize "in recognition of the extraordinary services he has rendered to the advancement of chemistry by his electrolytic theory of dissociation". Today this theory is one of the cornerstones of physical chemistry and electrochemistry.

Arrhenius later proposed definitions for acids and bases as an extension of his ionic theory. He also studied reaction rates as a function of temperature. In 1889 he introduced the concept of activation energy as the critical energy threshold required for molecules to react.

In 1905 Arrhenius became director of the Nobel Institute of Physical Chemistry. He died in Stockholm on October 2, 1927.

3

Solubility

3.1 Introduction

When a substance mixes with another substance to form a homogeneous mixture, a *true solution* is formed. The substance that is present in greater amount is called the *solvent* and the other substance is called the *solute*.

The solubility of a solid in a liquid is governed by many factors like nature of the solute, nature of the solvent, temperature, pressure, concentration of the solute, presence of other solutes, etc. The maximum amount of the solute that can dissolve in a fixed quantity of the solvent at a given temperature and pressure is called the *solubility* of the solute. Such a solution, which contains the maximum amount possible of the solute at that temperature and pressure, is called a *saturated solution*.

An ionic solid exists in the solid state as negative and positive ions packed into a regular, three-dimensional array called *crystal lattice*. The electrostatic attraction between the oppositely charged ions holds the crystal lattice together. The energy required to disband the ions from the lattice is called the *lattice energy* of the ionic solid. The lattice energy of a number of crystals is given in Table 3.1 in Appendix 3. For an ionic solid to dissolve in a solvent, the solvent has to overcome the attractive forces between the ions. This energy supplied by the solvent to overcome the force of attraction between the ions is called *solvation energy*. When the solvent involved is water, this energy is called *hydration energy*. The hydration energy of a number of crystals is given in Table 3.2 in Appendix 3.

When the solvation energy exceeds the lattice energy, the lattice breaks up and the individual ions drift into the solution. If the concentration of the ions is very low, they drift far apart from each other and there is virtually no interaction among them. Such a solution is called an *ideal-dilute* solution. However, if the concentration of the ions is high (more than 10^{-3}M), the ions are forced closer together. The electron clouds of adjacent ions interact with each other, leading to deviation from ideality. The effective concentration of the ions in solution is no longer equal to their actual concentration. The effective concentration of an ion in solution is known as its *activity*. The ratio of the activity of an ion to its actual concentration is called its *activity coefficient*. If the concentration of an ion is expressed in terms of its molality (moles of the ion per kilogram of the solvent) then its activity can be defined as

$$a = \gamma m \qquad (3.01)$$

where γ is the activity coefficient

m is the molality

In any solution of an ionic solid, cations and anions co-exist. There is no way to separate the contributions of the anions and the cations to the effective non-ideality of the solution. Therefore, the concept of a *mean ionic activity coefficient* is introduced to assign equal responsibility for the non-ideality to both kinds of ion. For a salt A_nB_m that dissolves to give n cations and m anions per formula unit, the mean ionic activity coefficient is given by

$$\gamma_\pm = (\gamma_+{}^n\gamma_-{}^m)^{1/s} \quad (3.02)$$

where $s = n + m$

γ_+ and γ_- are the activity coefficients of the cation and anion respectively.

3.2 The Debye-Hückel Limiting Law

In 1923, Peter Debye and Erich Hückel proposed an explanation for the departure from ideality in ionic solutions when the concentration of ions in the solution is high.

Debye and Hückel expressed the concentration of the ions in solution in terms of a quantity called *ionic strength*, defined as follows:

$$I = \tfrac{1}{2}\Sigma m_i z_i^2 \quad (3.03)$$

where m_i is the molality of the ith ion in solution

z_i is the charge on the ith ion

This expression emphasises the charge on the ions. Oppositely charged ions attract each other. At any given time, a positively charged ion is more likely to be surrounded by negatively charged ions, and vice-versa. Hence each ion in solution can be thought of as being surrounded by a "cloud" of oppositely charged ions, even though overall the solution is electrically neutral. The electrostatic interaction between the central ion and its surrounding cloud causes the deviation from ideality. At very low concentrations (less than 10^{-4} M), the mean ionic activity coefficient of the ions in solution can be related to the ionic strength of the solution by the *Debye-Hückel Limiting Law*:

$$\log \gamma_\pm = -\,|\,z_+ z_-\,|\,A\sqrt{I} \quad (3.04)$$

where $A = 0.509$ for an aqueous solution at 25°C

z_+ and z_- are the charges on the cation and anion respectively

I is the ionic strength of the solution.

3.3 The Extended Debye-Hückel Law

For solutions having higher concentrations, the mean ionic activity coefficient can be calculated using the *extended Debye-Hückel Law* :

$$\log \gamma_\pm = (-\,|\,z_+ z_-\,|\,A\sqrt{I})/(1 + a^\circ B\sqrt{I}) \quad (3.05)$$

where a° may be regarded as the effective diameter of the hydrated ions

$B = 0.33$ for an aqueous solution at 25°C.

3.4 Solubility Product

When the solvation energy supplied by the solvent is unable to overcome the lattice energy of an ionic solid, the solid remains practically undissolved. Such a salt is called sparingly soluble salt in that solvent. A sparingly soluble salt dissolves to such a small extent that it always forms a saturated solution. A dynamic equilibrium exists between the undissolved salt in the solid state and its dissociated ions in the solution.

Consider a sparingly soluble salt AB in solution.

$$AB_{(s)} \longleftrightarrow A^{+}_{(aq)} + B^{-}_{(aq)} \tag{3.06}$$

The equilibrium constant of this reaction is given by

$$K_{eq} = a_{A^+} + \frac{a_{B^-}}{a_{AB}} \tag{3.07}$$

where a_{A+}, a_{B-} and a_{AB} are the activities of A^+, B^- and AB respectively.

The activity of a solid is arbitrarily taken to be unity. Equation 3.07 can then be written as

$$K_{eq} = a_{A+} \cdot a_{B-} \tag{3.08}$$

The activity of the ions can be expressed in terms of their concentrations and activity coefficients.

$$K_{eq} = C_{A+} . a_{A+} \cdot C_{B-} a_{B-} \tag{3.09}$$

$$K_{eq} = K_c \cdot \gamma_{\pm}^{2} \tag{3.10}$$

where $K_c = C_{A+} \cdot C_{B-}$

$\gamma_{\pm}^{2} = \gamma_{+} \gamma_{-}$

The equilibrium constant in this case is known by a special name and symbol: the *true thermodynamic solubility product*, K_{sp}. K_c is the classical solubility product obtained by determining the actual concentrations of the ions in solutions. $\gamma_{\pm}$ can be determined from the extended Debye-Hückel Law.

$$K_{sp} = K_c \cdot \gamma_{\pm}^{2} \tag{3.11}$$

The solubility product of a few salts is given in Table 3.3 in Appendix 3.

3.5 Applications of Solubility Product

When the product of the ionic concentrations of a salt in a solution exceeds the value of its solubility product, the solute precipitates out from the solution. This property is used to separate ionic salts from mixtures. In the qualitative analysis of cations, the division of cations into six groups is based on the value of their solubility products.

3.6 Heat of Solution

The dissolution of a solute in a solvent is accompanied by a heat change. Thermochemistry is the branch of chemistry that deals with the study of heat changes accompanying a chemical reaction.

A chemical reaction, during which heat is lost by the reacting system and gained by the surroundings, is called an exothermic reaction. On the other hand, if, during the reaction, heat is absorbed by the reacting system and lost by the surroundings, the reaction is called endothermic. When the heat change is measured at constant pressure, it is called the enthalpy change ΔH of the reaction. Enthalpy changes during a reaction are generally expressed for a set of standard conditions. The standard state of a substance is the most stable state of the substance at one atmosphere pressure and temperature 298K.

The state of the solution is described by specifying the quantities of both solvent and solute. It is necessary to specify the state of the solution when describing the enthalpy change on dissolution because the amount of heat evolved or absorbed depends upon the composition of the solution.

The *integral enthalpy of solution* at a given concentration is defined as the change in enthalpy when one mole of the solute in its standard state is completely dissolved in enough of the solvent to produce a solution having the desired composition.

The *differential enthalpy of solution* is the enthalpy change when one mole of the solute is dissolved in such a large quantity of the solution that any concentration change is negligible. This quantity is also called the *heat of solution.* It can be alternatively defined as the heat change when one mole of the solute is dissolved in a solution that is nearing saturation point.

If the process of dissolution is endothermic, the solubility of the solute increases with temperature. On the other hand, if the process of dissolution is exothermic, the solubility of the solute decreases with temperature. In most cases, it has been observed that the solubility of a substance rises with temperature. The variation in solubility with temperature of a few salts is shown in Figure 3.1.

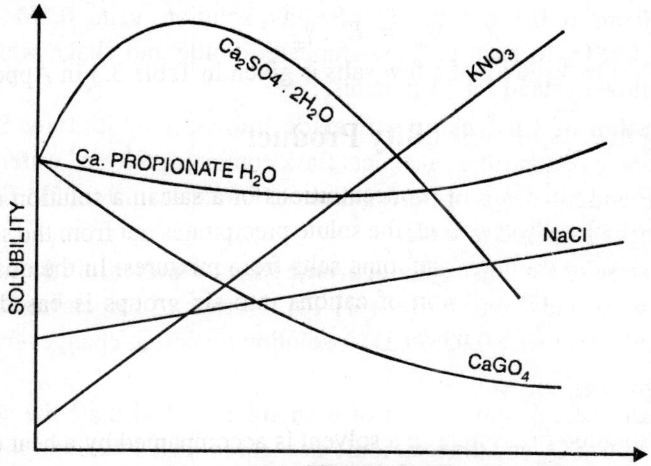

Figure 3.1: The variation in solubility of a few salts with temperature. (Diagram not to scale.)

The quantitative relationship between solubility and temperature is given by the van't Hoff isochore.

$$(\partial \ln S/\partial T)_P = \Delta H/RT^2 \quad (3.12)$$

where S is the solubility of the solute
ΔH is the heat of the solution
R is the universal gas constant
T is the absolute temperature

Integrating the above equation,

$$\ln S_1 - \ln S_2 = \Delta H(T_2 - T_1)/RT_1T_2 \quad (3.13a)$$

or

$$\ln S = -\Delta H/RT + \text{constant} \quad (3.13b)$$

Thus, the heat of solution can be determined by measuring the solubility at different temperatures. The above integration assumes that the heat of solution is independent of temperature, which is not strictly true.

EXPERIMENTS

Experiment 3.1

Aim: To determine from solubility measurements the true thermodynamic solubility product of calcium sulphate at room temperature.

Requirements

Calcium sulphate, 0.5M KNO_3 0.1M NaOH, standard 0.05M EDTA, murexide or Eriochrome Black T indicator.

Procedure

Prepare 100 cm^3 of 0.1, 0.2, 0.3, 0.4M KNO_3 solutions using 0.5M solution. Add 2g of $CaSO_4$ to each in glass-stoppered bottle and shake well for 45 minutes. Allow to stand for 45 minutes.

Estimation of Ca^{2+} using murexide indicator: Withdraw 5 cm^3 of solution using a cotton filter in conical flask containing 90 cm^3 water, 10 cm^3 0.2N NaOH and 5-6 drops of fresh indicator. Titrate against 0.05M EDTA till colour changes from red to violet.

Estimation of Ca^{2+} using Eriochrome Black T indicator: 5 cm^3 analyte + 90 cm^3 water and 2 cm^3 buffer (pH 10), 2 drops of Eriochrome Black T indicator and titrate with 0.05M EDTA solution till colour changes from wine red to blue.

Calculate the calcium content of each solution. Calculate the solubility product K_c of $CaSO_4$ and the total ionic strength in each solution. Plot a graph of pK_c against $\sqrt{I}/(1+1.25\sqrt{I})$ and determine K_s from the y-intercept. Finally calculate $\gamma_\pm$ at each concentration from the relation $K_s = K_c \cdot \gamma_\pm^2$

Observations and calculations

Standardisation of 0.05M (approx) EDTA solution

Solution in burette: 0.05M (approx) EDTA

Solution in conical flask: 10 cm^3 0.05M $CaCl_2$ + 10 cm^3 0.2N NaOH

Indicator: murexide

End point: red to violet

$$EDTA \equiv CaCl_2$$
$$N_1V_1 = N_2V_2$$
$$N_1 = \_\_\_\_ \text{ N}$$

Conc. of KNO_3 (M)	Vol of 0.05M EDTA V_1 (cm^3)	C_{Ca2+} = $0.05V_1/5$ (M)	K_c = C^2_{Ca2+}	pK_c = $-\log K_c$	I_{CaSO4} = $4C_{Ca2+}$

I_{KNO_3} = C_{KNO_3}	Total I = $I_{KNO_3} + I_{CaSO_4}$	$\sqrt{I}$	$1.25\sqrt{I}$	$\sqrt{I}$ / $(1+1.25\sqrt{I})$	$-2.036\sqrt{I}$ / $(1+1.25\sqrt{I})$

$2 \log \gamma_\pm$ = $2[-2.036\sqrt{I}/(1+1.25\sqrt{I})]$	pK_s = $pK_c - 2\log\gamma_\pm$	K_s = antilog $(-pK_s)$	$\gamma_\pm$ = $\sqrt{(K_s/K_c)}$

Mean K_s = ____

From the graph of pK_c against $\sqrt{I}/(1+1.25\sqrt{I})$, pK_s = y-intercept = ____

Therefore K_s = antilog $(-pK_s)$ = ____

Result

1. The approximate solubility product of $CaSO_4$ = K_c = ____ (mol $dm^{-3})^2$ (by calculation)
2. The true thermodynamic solubility product of $CaSO_4$ = K_s = ____ (by graph)

Conc. of KNO_3 (M)	$\gamma_\pm = \sqrt{(K_s/K_c)}$

Experiment 3.2

Aim: To study the variation in the solubility of potassium hydrogen tartrate with ionic strength and hence determine the mean ionic activity coefficient.

Theory

Potassium hydrogen tartrate is a sparingly soluble salt. When dissolved in a solution that contains an excess of potassium ions, its solubility gets suppressed due to the common ion effect. The solubility product of potassium hydrogen tartrate can be expressed as

$$K_s = (c_{\pm}\gamma_{\pm})^2 \tag{3.14}$$

where $c_{\pm}$ is the concentration of potassium hydrogen tartrate and $\gamma_{\pm}$ is the mean ionic activity coefficient of its ions. $\gamma_{\pm}$ can then be expressed as

$$\gamma_{\pm} = (K_s)^{1/2} / c_{\pm}$$

$$\log \gamma_{\pm} = \tfrac{1}{2} \log K_s - \log c_{\pm} \tag{3.15}$$

According to the Debye-Hickel Limiting Law (equation (3.04)),

$$\log \gamma_{\pm} = -\,|\,z_+ z_-\,|\, A \sqrt{I}$$

Equating (3.15) and (3.04),

$$\tfrac{1}{2} \log K_s - \log c_{\pm} = -\,|\,z_+ z_-\,|\, A \sqrt{I} \tag{3.16a}$$

Rearranging this equation,

$$\log c_{\pm} = -\,|\,z_+ z_-\,|\, A \sqrt{I} + \tfrac{1}{2} \log K_s \tag{3.16b}$$

Thus a graph of log $c_{\pm}$ against $\sqrt{I}$ will be linear with a y-intercept of ½ log K_s. Knowing K_s, $\gamma_{\pm}$ can be calculated at any concentration by substituting in equation (3.15).

Requirements

Stoppered bottles, thermostat at 25 °C, KCl, NaCl, potassium hydrogen tartrate, standard 0.01M NaOH

Procedure

Prepare the following series of solutions (100 cm^3 each):

Set I: KCl solution: 0.025, 0.05, 0.1, 0.2, 0.3M

Set II: NaCl + KCl mixture containing the same amount of KCl as in Set I but made up with NaCl so that the ionic strength is 0.2 mol dm^{-3}

Add excess solid potassium hydrogen tartrate to each solution in a glass-stoppered bottle and shake in a thermostat at 25°C for 1-2 days. Filter each solution through a dry filter paper, rejecting the first 10 cm^3 of filtrate.

Titrate 10 cm^3 sample of each with the standard NaOH solution using phenolphthalein as indicator. From the titrations, determine the concentration of tartrate c_2 mol dm^{-3} in the solution and hence

$$c_{\pm} = \sqrt{[c_1 (c_1 + c_2)]}$$

where c_1 mol dm^{-3} is the concentration of KCl in solution.

Plot $\log c_\pm$ against $\sqrt{I}$ taking into account the potassium hydrogen tartrate also. Extrapolate the line, determine K_s and hence $\gamma_\pm$ at each ionic strength. For both series, plot $\log \gamma_\pm$ against $\sqrt{I}$.

Observations and calculations

For Sets I and II

Conc. of KCl c_1 (M)	Vol of 0.01N NaOH V_1 (cm^3)	Conc. of tartrate (M) $c_2 = 0.01V_1/10$	$c_\pm = \nu\,[c_1\,(c_1 + c_2)]$	$\log c_\pm$

$I = I_1 + I_2 = c_1 + c_2$ (M)	$\sqrt{I}$	$\gamma_\pm = (\sqrt{K_s})/c_\pm$	$\log \gamma_\pm$

From the graph of $\log c_\pm$ against $\sqrt{I}$, y-intercept = ½ $\log K_s$ = ______

Therefore, K_s = ______

Result

For Sets I and II,

Conc. of KCl (M)	Mean ionic activity coefficient of potassium hydrogen tartrate $\gamma_\pm$

Experiment 3.3

Aim: To study the variation in the solubility of calcium hydroxide in the presence of sodium hydroxide, and hence determine the solubility product of calcium hydroxide at room temperature.

Theory

In a saturated solution there is dynamic equilibrium between two opposing reaction proceeding at the same time. In the case of calcium hydroxide, the reaction is solution can be represented as

$$Ca(OH)_2 \longleftrightarrow Ca^{2+} + 2OH^-$$

The rate at which calcium hydroxide dissociates into its component ions is the same as the rate at which calcium ions and hydroxide ions precipitate out from the solution as solid calcium hydroxide. Therefore the classical solubility product of calcium hydroxide can be determined by the product of the concentrations of ions in the solution.

$$K_c = C_{Ca2+} \cdot C^2_{OH-}$$

where C_{Ca2+} and C_{OH-} are the concentrations in mol dm^{-3} of Ca^{2+} ions and OH^- ions respectively in saturated solution.

When sodium hydroxide is added to the solution, OH^- ions become in excess. This causes the OH^- concentration to become very large. To maintain the constant value of the solubility product, the equilibrium shifts to the left. Hence the concentration of Ca^{2+} decreases and so does the solubility of calcium hydroxide.

Requirements

0.1M sodium hydroxide solution, 0.05N HCl solution, pure solid calcium hydroxide, phenolphthalein indicator.

Procedure

In four stoppered 250 ml bottles, prepare the following mixtures:

Bottle no.	Volume of (cm^3)		Calcium hydroxide (g)
	0.1N NaOH	Water	
I	100	100	2
II	50	150	2
III	25	175	2
IV	0	200	2

Stopper the flasks tightly and shake them vigorously for half an hour. Allow the mixtures to stand for an hour. Plug the end of a 20 cm^3 pipette with cotton wool. Pipette out 20 cm^3 from each mixture and titrate against standard HCl using methyl orange as indicator. The end point is from yellow to orange.

The concentration of OH^- ions obtained by titration is the total concentration due to NaOH and $Ca(OH)_2$. The concentration of OH^- ions due to $Ca(OH)_2$ alone can be obtained by subtracting the initial concentration of OH^- ions due to NaOH from the total concentration. Since every molecule of $Ca(OH)_2$ on dissociation gives two OH^- ions and one Ca^{2+} ion, the concentration of Ca^{2+} ions is half the concentration of OH^- ions. Knowing the concentration of Ca^{2+} and OH^- ions, the solubility and hence the solubility product of calcium hydroxide can be determined.

Observations and calculations

Titration

Solution in burette: 0.05M HCl

Solution in flask: 20 cm^3 mixture

Indicator used: methyl orange

End point: yellow to orange

Bottle no.	Titre reading V_{HCl} (cm^3)	Total C_{OH^-} $= C_{NaOH} + C_{Ca(OH)2}$ $= C_{HCl} \times V_{HCl} / 20$ B (M)	Initial C_{OH^-} $= C_{NaOH}$ A (M)
I ... IV			

Bottle no.	C_{OH^-} due to $Ca(OH)_2$ $B - A$ (M)	$C_{Ca^{2+}} =$ ½ $(B - A)$ (M)	Solubility product K_c = ½ $(B - A)^3$ (mol dm^{-3})^3
I ... IV			

Mean K_c = ________ (mol dm^{-3})^3

Result

The solubility of calcium hydroxide __________(increases/decreases/remains the same) with increasing concentration of sodium hydroxide in the solution.

Solubility product of calcium hydroxide = __________ (mol dm^{-3})^3

Experiment 3.4

Aim: Determination of the solubility of benzoic acid (or salicylic acid) over a range of temperatures and hence the heat of solution.

Requirements

Thermostat, conical flasks, 10 cm^3 pipette connected to an extension tube packed with glass wool by rubber tubing, pure benzoic acid and salicylic acid, 0.05M NaOH solution, 0.05M succinic acid solution

Procedure

Prepare about 200 cm^3 of solution of benzoic acid saturated at about 60°C and transfer 30 – 40 cm^3 into each conical flask. Place a conical flask in the thermostat maintained at the required temperature for 30 minutes and stir well while maintaining the temperature. Immediately withdraw 10 cm^3 sample, drain into a weighed conical flask and reweigh. Under no circumstances must any small crystals, which have separated out on cooling, be withdrawn. This can be avoided by fitting to the pipette, with rubber tubing, a piece of drawn-out tube packed with glass wool. Remove the end piece before discharging into the conical flask. Titrate with 0.05M NaOH solution using phenolphthalein as indicator. Standardise the sodium hydroxide solution using succinic acid.

Make duplicate measurements of solubility at 40°C, 35°C and 25°C. Repeat for salicylic acid.

Determine the solubility of the acid S, expressed as mol kg^{-1} at each temperature. Plot a graph of log S against $1/T$ and determine the heat of solution.

Observations and calculations

Molecular mass of salicylic acid = 138.1 g mol^{-1}

Molecular mass of benzoic acid = 122.1 g mol^{-1}

Concentration of NaOH used = A = ___ N

Temp T (°C)	Wt of empty flask w_1 (g)	Wt. of flask & acid w_2 (g)	Wt. of 10 cm^3 of acid $w_2 - w_1$ (g)	Vol. of NaOH required for 10 cm^3 acid V_1 (cm^3)	Wt. of acid w_3 (g)
25					
35					
40					

Calculation for mass of acid w_3

1000 cm^3 of 1 N NaOH ≡ equivalent wt. of acid

Now, equivalent wt. of acid = Mol wt/basicity of acid

Since benzoic and salicylic acids are both monobasic acids,

1000 cm^3 of 1 N NaOH ≡ Mol wt of acid

Therefore, V_1 (cm^3) of A N NaOH ≡ Mol wt of acid (V_1 A/1000)

w_3 = Mol wt of acid (V_1 A/1000)

Wt of water $(w_2 - w_1) - w_3$ (g)	Solubility S in g kg^{-1} of the solvent	Solubility S in mol kg^{-1} of the solvent	log S	Temp T (K)	1/T (K^{-1})

Calculation for solubility S

$[(w_2 - w_1) - w_3]$ g of water contains w_3 g of acid.

Therefore, 1000 g of water contains 1000 $w_3/[(w_2 - w_1) - w_3]$ g of acid.

Solubility S = 1000 $w_3/[(w_2 - w_1) - w_3]$ g kg^{-1} of the solvent

= 1000 $w_3/M[(w_2 - w_1) - w_3]$ mol kg^{-1} of the solvent

where M is the Mol wt of the acid

Calculation for ΔH

From the graph of log S versus $1/T$,

Slope = _______

From the Van't Hoff isochore,

$$\text{Slope} = -\Delta H/2.303R$$

Therefore, $\Delta H = -\text{slope} \times 2.303R$ = _______ kJ mol^{-1}

Result

Heat of solution of benzoic/salicylic acid = _______ kJ mol^{-1}

Appendix

Review Questions

1. Does solubility depend upon particle size? If so, how?
2. Rubber can dissolve in benzene but not in water. Why?

Table 3.1: Lattice Enthalpy (kJ mol^{-1}) for the reaction $MX_{(s)} \rightarrow M^+_{(g)} + X^-_{(g)}$

Cations	*Anions*					
	F^-	Cl^-	Br^-	I^-	O^{2-}	S^{2-}
Li^+	1037	852	815	761	–	2472
Na^+	926	787	752	705	–	2203
K^+	821	717	689	649	–	2052
Rb^+	789	695	668	632	–	1949
Cs^+	750	676	654	620	–	1850
Be^{2+}	3505	3017	2914	2800	–	–
Mg^{2+}	2957	2524	2440	2327	3850	3406
Ca^{2+}	2630	2255	2176	2074	3461	3119
Sr^{2+}	2492	2153	2075	1963	3283	2974
Ba^{2+}	2352	2056	1985	1877	3114	2832
Ag^+	969	912	900	886	–	2677

Table 3.2: Standard Hydration Enthalpy at infinite dilution for the reaction $X^+_{(g)} + Y^-_{(g)} \rightarrow X^+_{(aq)} + Y^-_{(aq)}$

	F^-	Cl^-	Br^-	I^-
Li^+	–1026	–884	–856	–815
Na^+	–911	–783	–742	–701
K^+	–828	–685	–658	–617
Rb^+	–806	–664	–637	–613
Cs^+	–782	–640	–613	–572

Table 3.3: Thermodynamic Solubility Product K_{sp}

Salt	Solubility Product	Salt	Solubility Product
$AgBrO_3$	5.34×10^{-5}	$FeCO_3$	3.07×10^{-11}
AgBr	5.35×10^{-13}	$Fe(OH)_2$	4.87×10^{-17}
AgCl	1.77×10^{-10}	FeS	1.59×10^{-19}
Ag_2CrO_4	1.12×10^{-12}	$Fe(OH)_3$	2.64×10^{-39}
AgI	8.51×10^{-17}	Li_2CO_3	8.15×10^{-4}
Ag_2S (α)	6.69×10^{-50}	$MgCO_3$	6.82×10^{-6}
Ag_2S (β)	1.09×10^{-49}	$Mg(OH)_2$	5.61×10^{-12}
$BaCO_3$	2.58×10^{-9}	$MgC_2O_4.2H_2O$	4.83×10^{-6}
CdS	1.40×10^{-29}	$Mg_3(PO_4)_2$	9.86×10^{-25}
$CaCO_3$	4.96×10^{-9}	MnS	4.65×10^{-4}
$Ca(OH)_2$	4.68×10^{-6}	NiS	1.07×10^{-21}
$CaC_2O_4.2H_2O$	2.34×10^{-9}	$PbBr_2$	6.60×10^{-6}
$Ca_3(PO_4)_2$	2.07×10^{-33}	$PbCO_3$	1.46×10^{-13}
$CaSO_4$	7.10×10^{-5}	$PbCl_2$	1.17×10^{-5}
CuBr	6.27×10^{-9}	PbC_2O_4	8.51×10^{-10}
CuCl	1.72×10^{-7}	$PbSO_4$	1.82×10^{-8}
CuI	1.27×10^{-12}	PbS	9.04×10^{-29}
Cu_2S	2.26×10^{-48}	$SrCO_3$	5.60×10^{-10}
CuS	1.27×10^{-36}	ZnS	2.93×10^{-25}

Table 3.4: Solubilities of a few commonly used reagents at 293 K

Reagent	Formula	Solubility / (g per 100 cm^3 of water)
Ammonium chloride	NH_4Cl	29.1
Ammonium nitrate	NH_4NO_3	86.3
Ammonium sulphate	$(NH_4)_2SO_4$	53.5
Potassium chloride	KCl	29.8
Potassium hydroxide	KOH	81.3
Sodium acetate	CH_3COONa	46.5
Sodium carbonate	Na_2CO_3	20.9
Sodium chloride	NaCl	31.6
Sodium hydroxide	NaOH	80.3

Expected Experimental Results

Experiment 3.1

The true thermodynamic solubility product of $CaSO_4$ is 7.10×10^{-5}.

Experiment 3.2

For Sets I and II, it is observed that the mean ionic activity coefficient of potassium hydrogen tartrate decreases with increasing concentration of potassium chloride.

Experiment 3.3

The solubility of calcium hydroxide decreases with increasing concentration of sodium hydroxide in the solution.

Thermodynamic solubility product of calcium hydroxide is 4.68×10^{-6}.

Experiment 3.4

Heat of solution of benzoic acid is approximately 25 kJ mol^{-1}.

Heat of solution of salicylic acid is approximately 37 kJ mol^{-1}.

Hints to Review Questions

1. Solubility does not depend upon particle size. However, the rate at which a substance dissolves does depend upon particle size. For example, powdered sugar dissolves much faster in cold water than granular sugar.
2. The solubility of an amorphous polymer like rubber in a low molar mass solvent like water or benzene can be estimated by employing the concept of solubility parameter. The solubility parameter reflects the strength of the intermolecular forces between the molecules of the solvent. These intermolecular forces may be dispersion forces, dipole forces and hydrogen bonding. For an amorphous polymer to be soluble in a low molar mass solvent, the difference between their solubility parameters must be fairly narrow. The solubility parameters of rubber, benzene and water are 8.05, 9.05 and 23.43 $(\text{cal cm}^{-3})^{1/2}$ respectively.

Further Exploration through the Internet

http://www.hyper-ad.com/tutoring/chemistry/Sol_prod.html

http://ir.chem.cmu.edu/irproject/curriculum/conceptTests/solubility/

Gilbert Newton Lewis

G N Lewis was born on October 23, 1875 in Weymouth, USA. His doctoral work was on the thermodynamics of zinc and cadmium amalgams and he was awarded his Ph. D. from Harvard University in 1899. Like many chemists of his time, he then went to Europe. There he worked with Ostwald in Leipzig and Nernst in Guttingen. He returned to Harvard in 1901. In 1905, he joined the research group working under A. A. Noyes at the Massachusetts Institute of Technology.

At that time, the law of conservation of energy and other thermodynamic relations were known only as isolated equations. Lewis built on the work of J. W. Gibbs to organise the field of chemical thermodynamics and apply its laws to real chemical systems. In 1907, Lewis introduced the concept of partial molar quantities, which was of immense help in the quantitative study of solutions. His work in thermodynamics led him to confirm Nernst's Third Law of Thermodynamics. Lewis also undertook the systematic determination of the electrode potential, conductivity, free energy and other thermodynamic constants of the elements. These tables are in use even today.

In 1912, Lewis was appointed the chairman of the Department of Chemistry and the Dean of the College of Chemistry at the University of California, Berkeley. In those days, there was no serious science in California and his first labs were in a temporary wooden building. However, by the end of his 34-year tenure, the Berkeley Chemistry Department had transformed into one of the finest research centres in the world. Although Lewis never received the Nobel Prize, five of his students eventually did: H. Urey (1934), W. Giauque (1949), G. Seaborg (1951), W. Libby (1960) and M. Calvin (1961).

Lewis is best remembered for his development of the valence bond theory and the electron pair bonding diagrams, which he first published in 1916. In 1923, he redefined acids and bases in terms of the octet rule. His book *Valence and the Structure of Atoms and Molecules* is one of the greatest contributions to the modern theory of bonding. In 1923 was also published his most famous book *Thermodynamics and the Free Energy of Chemical Substances*, which he co-authored with Merle Randall.

Lewis' later research on isotopes predicted the existence of naturally occurring heavy water. In 1933, he prepared a pure sample of heavy water. Other areas of his interest included neutron physics, natural radioactivity and photochemistry.

On March 23, 1946 Lewis died unexpectedly in his laboratory. Surrounded by his beakers and test tubes, he was carrying out an experiment on fluorescence. For a man who once said "Physical Chemistry is everything that is interesting." it was perhaps the best way to go.

4

Partial Molar Properties

4.1 Introduction

When a gas is filled inside a container, the molecules of the gas collide with the walls of the container due to their ceaseless random motion. This exerts a pressure on the walls of the container. When a mixture of gases is filled inside a container, each gas component makes a certain contribution to the total pressure exerted. In the nineteenth century, John Dalton observed that the pressure exerted by a mixture of gases is the sum of the pressures that each gas would exert if it alone occupied the container. The pressure exerted by a component gas is called its *partial pressure*. If a gaseous mixture consisted of n_A moles of gas A and n_B moles of gas B then the total pressure p exerted by the mixture would be

$$p = p_A + p_B \tag{4.1}$$

where p_A and p_B are the partial pressures of gases A and B.

The *mole fraction* x_A of gas A is the ratio of the number of moles of A to the total number of moles of gas present in the mixture.

$$x_A = n_A/(n_A + n_B) \tag{4.2}$$

The partial pressure of a gas can be expressed in terms of its mole fraction.

$$p_A = x_A\, p \tag{4.3}$$

where p is the total pressure of the mixture.

The properties of mixtures of gases can be discussed in terms of the partial pressures of their components. A more general discussion of the thermodynamics of mixtures needs the introduction of other analogous partial properties.

4.2 Partial Molar Properties

To define any system fully, we must know its composition c, volume V, pressure p, and temperature T. These four variables are of fundamental importance and are called *thermodynamic variables* or *state variables*. Some of the state variables are proportional to the size of the system. For example, a bulky system has more volume than a compact one. Such variables are called *extensive* variables. Mass m, volume V, and the number of moles present n, are some examples of extensive variables. Other variables do not depend on the size of the system and are called *intensive* variables. For example, joining two systems that are at the same temperature would not double the temperature of the resulting system. Pressure p, temperature T, concentration c, and mole fraction x, are some examples of intensive variables.

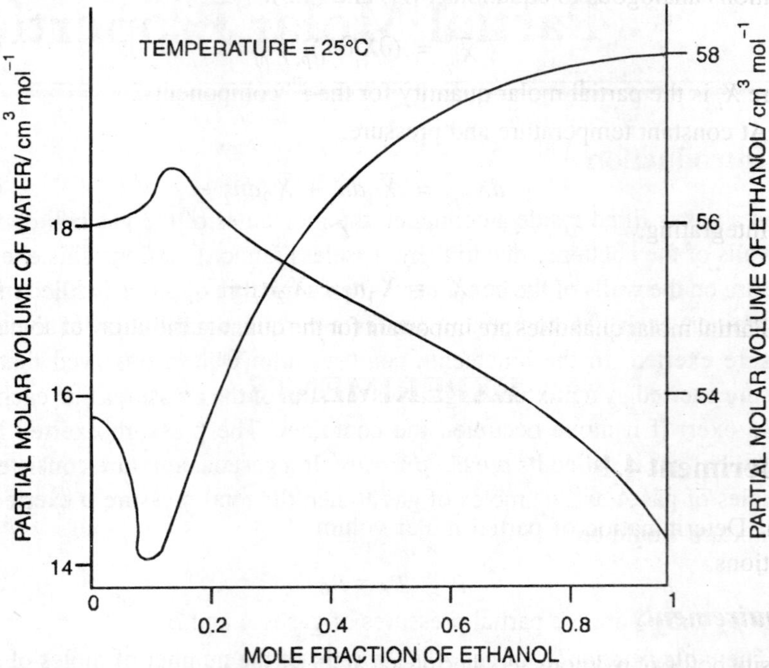

Figure 4.1: Variation in the partial molar volume of water and ethanol with mole fraction of ethanol.

In systems whose composition can change due to the addition or removal of substances, any extensive property of the system is a function of the number of moles of the components, in addition to temperature and pressure. In other words, the extensive properties of a system are not truly additive when pure components are mixed.

For example, when 1 mol of water at 25°C is added to a huge volume of pure water, the volume increases by 18 cm^3. Hence, 18 $cm^3\ mol^{-1}$ is the molar volume of pure water. However, when 1 mol of water is added to a huge volume of pure ethanol, the increase in volume is only 14 cm^3. The quantity 14 $cm^3\ mol^{-1}$ is said to be the partial molar volume of water in pure ethanol.

The *partial molar volume* of a substance can thus be defined as the change in volume when one mole of the substance is added to a very large volume of the mixture. Mathematically, it is expressed as

$$\overline{V}_A = (\partial V/\partial n_A)_{p,\,T,\,nB} \tag{4.4}$$

The total volume of a mixture can be defined in terms of the partial molar volumes of its components in an equation analogous to equation (4.1).

$$\overline{V} = n_A\overline{V}_A + n_B\overline{V}_B \tag{4.5}$$

Diagram: partial molar volumes of water and ethanol

In general, any partial molar property X can be expressed in terms of equations analogous to equations (4.4) and (4.5).

$$\overline{X}_i = (\partial X/\partial n_i)_{p,\,T,\,nj} \tag{4.6}$$

where X_i is the partial molar quantity for the ith component.

At constant temperature and pressure,

$$dX_{T,p} = \overline{X}_1 dn_1 + \overline{X}_2 dn_2 + \ldots \tag{4.7}$$

Integrating,

$$X = \overline{X}_1 n_1 + \overline{X}_2 n_2 + \ldots \tag{4.8}$$

Partial molar quantities are important for the quantitative study of solutions.

EXPERIMENTS

Experiment 4.1

Aim: Determination of partial molar volume of methanol in dilute aqueous solutions.

Requirements

Density bottle or pyknometer, thermostat at 25.00 ± 0.05 °C, absolute methanol.

Procedure

The density bottle is used for accurate determination of density of liquids. The density bottle should be cleaned and dried, by washing well with acetone followed by distilled water and then again by acetone. The density bottle is first weighed empty. It is then filled up to the mark with distilled water and weighed again. Let the weight of water be W. If the volume of the density bottle is V then the density of distilled water is W/V. The density bottle is then cleaned and dried and filled with pure methanol. The bottle is weighed. Let the weight of methanol be W'. The density of methanol is W'/V.

Determine the density of pure alcohol very accurately by using the method described above. Prepare 20 g of mixtures of water and methanol by taking 20, 40, 60, 80% (wt/wt) of methanol. Determine the density of each of these solutions as well.

Calculate the specific volume $1/\sigma$ of each solution, including that of pure methanol and distilled water. Plot a graph of $1/\sigma$ against percentage by weight of methanol. Draw a smooth curve through the points and draw tangents to this curve at different concentrations. Extrapolate the tangents till they intercept on the ordinate at 0% and 100% weight of methanol. The values at the two intercepts give the partial specific volume for water and methanol respectively at that concentration.

Calculate the partial molar volume using the formula:

Partial molar volume = partial specific volume × relative molecular mass

Calculate the mole fraction of methanol in each mixture. Draw a curve showing the variation of the partial molar volume of methanol with mole fraction of methanol. Draw a curve showing the variation of the partial molar volume of water with mole fraction of methanol.

Observations and Calculations

Mass of empty density bottle w_1 = _____ g
Volume of density bottle V = _____ cm^3
Density of pure methanol = _____ g cm^{-3}
Density of pure water = _____ g cm^{-3}
Relative molecular mass of methanol = 32.042 g mol^{-1}
Relative molecular mass of water = 18.015 g mol^{-1}

% mass of methanol	Volume of methanol (cm^3)	Vol. of water (cm^3)	Mole fraction of methanol x_A	Wt. of liquid + bottle w_2 (g)	Wt. of liquid $w_2 - w_1$ (g)	Density of mixture σ = $(w_2 - w_1)$/ vol of density bottle (g cm^{-3})	Specific Volume 1/σ (g^{-1} cm^3)
0							
20							
40							
60							
80							
100							

% wt. of CH_3OH	Partial specific vol. of water (cm^3)	Partial specific vol. of CH_3OH (cm^3)	Partial molar volume of water (cm^3 mol^{-1})	Partial molar volume of CH_3OH (cm^3 mol^{-1})
0				
20				
40				
60				
80				
100				

Result

1. Partial molar volume of water in pure methanol = _____ cm^3 mol^{-1}
2. Partial molar volume of methanol in pure water = _____ cm^3 mol^{-1}

Appendix

Review Questions

1. What is the physical significance of partial molar properties?
2. Do biological reactions take place in open or closed systems?

Expected Experimental Results

Partial molar volume of methanol in pure water is 38.7 cm^3 mol^{-1} at 25 °C.

Hints to Review Questions

1. The partial molar property of a substance represents the increase in the property of the system when, at constant temperature and pressure, one mole of that substance is added to such a large quantity of the system that the composition of the system remains virtually unchanged. The partial molar property of a constituent of a system can also be regarded as the contribution of one mole of that constituent to the total value of the property of the system under the specified conditions. In general, the value of a partial molar property of a constituent in a solution is not equal to the value for the pure substance. In fact, the value of a partial molar property of a constituent in a solution varies with the composition of the solution. The partial molar property is independent of the number of moles of a constituent in the system; it depends only on the composition of the system at a given temperature and pressure.
2. Open systems exchange both energy and matter with the surroundings. Therefore, an open system is not at equilibrium with its surroundings, unlike closed systems. Many processes fundamental to biology and life are open, non-equilibrium and non-linear systems, extracting energy from their surroundings for their existence.

Further Exploration through the Internet

http://www.chem.arizona.edu/~salzmanr/480a/480ants/mixpmqis/mixpmqis.html

http://chemeng.stanford.edu/~charles/mse202/assets/multimedia/Chap8.main.pdf

http://web.mit.edu/10.40/www/pdf/ch9notes.pdf

http://www.ubishops.ca/ccc/div/sci/chem/223a/L024.html

http://www.chem.vt.edu/chem-dept/schug/3615/mx.html

Friedrich Wilhelm Ostwald

Wilhelm Ostwald was born on September 2, 1853 in Riga, Latvia. He did his doctorate from the University of Dorpat in 1878. He taught at Riga before going to the University of Leipzig in 1887. He started his experimental research in 1875 with studies on the dynamics of reactions in solution. He then turned to the field of electrochemistry and discovered the law of dilution, which is named after him. He also studied the electrical conductivity and electrolytic dissociation of organic acids. He invented a viscometer that is still used for measuring the viscosity of solutions.

Ostwald was a prolific writer and had a gift for organisation. He used these talents to almost single-handedly establish physical chemistry as a nearly independent branch of chemistry. He published numerous textbooks and was chiefly responsible for the founding of the *Zeitschraft fur Physikalische Chemie* (Journal of Physical Chemistry) in 1887. The journal was for long the most influential publication in the field and Ostwald himself edited 100 volumes up to 1922.

In 1894 Ostwald turned his attention to catalytic reactions and gave the first modern definition of a catalyst. In 1900 he devised a method to manufacture nitric acid by oxidising ammonia. This process was named after him and is of great industrial importance even today. In 1909 he was awarded the Nobel Prize "in recognition of his work on catalysis and for his investigations into the fundamental principles governing chemical equilibria and rates of reaction".

Ostwald remained interested in research right up to the end of his life. He extensively studied the phenomenon of colour. He also wrote on the philosophy of science and took active part in public life.

Ostwald died at his country home near Leipzig on April 3, 1932.

5

Reaction Kinetics

5.1 Introduction

Chemical reactions vary widely in the amount of time they require to go to completion. The precipitation of an insoluble salt from its constituent ions in aqueous solution takes place in a fraction of a second, while an iron nail left in the open air would take months to start rusting. The conversion of the carbon content of dead organic matter into petroleum at the bottom of the sea takes millions of years. Some reactions never go to completion; they stop at some intermediate, equilibrium position.

Chemical kinetics is the study of the rates of reactions and the various factors that influence them. The *rate* of a chemical reaction is the amount of reaction that occurs in unit time. To measure this rate, we choose a property of the reaction that can be observed without affecting the reaction itself. For example, in a reaction where a colourless substance is changing to a coloured one, the rate at which the intensity of the colour increases would give the rate of the reaction. Other physical methods measure changes in volume, pressure, optical rotation, electrical conductivity, refractive index, pH or viscosity (especially for polymerisation reactions). It is also possible to study reaction kinetics by chemical analysis. The reaction is stopped after a definite time by cooling or by the addition of some reagent. The reaction mixture is then analysed by means of some chemical reaction.

According to elementary kinetic theory, a chemical reaction will occur only when the molecules of the reacting substances collide with each other. However, not all collisions result in a reaction. Only those molecules that possess energy equal to or greater than a certain threshold value will react in a collision to give rise to the product. This threshold energy is called *activation energy* E_a, and is characteristic of each reaction.

The variation in the energy of the reactants and products as the reaction proceeds is shown in Figure 5.1.

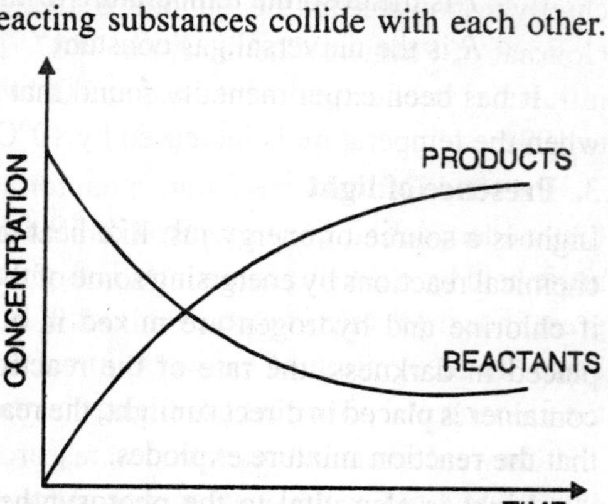

Figure 5.1: Variation in the concentration of reactants and products with time.

The activation energy acts as an energy barrier that the reactants must overcome if the reaction is to proceed.

5.2 Factors Affecting the Rate of a Reaction

A number of factors affect the rate of a reaction. Some of these are concentration of the reactants, temperature of the reaction mixture, the pressure of the reactants (in case of gases), the surface area of the reactants, the presence of light and the presence of a catalyst. We will study these in brief:

1. Concentration of the reactants

A chemical reaction occurs when the molecules of the reactants collide with each other. The rate of the reaction therefore depends on the frequency with which the molecules collide, which in turn depends on their concentration. The more 'crowded' the particles are, the more often they will bump into each other, thus leading to an increase in the rate of the reaction.

2. Temperature of the reaction mixture

When the temperature of a reaction is increased, heat is supplied to the molecules of the reactants. The molecules acquire greater kinetic energy and hence collide with each other more often. This leads to an increase in the rate of the reaction.

Arrhenius quantified the relationship between the reaction rate and the temperature through his equation

$$k = A e^{-(Ea/RT)} \tag{5.1}$$

where k is the rate constant of the reaction (the rate of the reaction when the concentrations of all reactants is unity)

A is a constant known as frequency factor. It has the same units as the rate constant.

E_a is the activation energy in J

T is the absolute temperature of the reaction in degree kelvin

R is the universal gas constant

It has been experimentally found that the rates of most reactions double when the temperature is increased by 10°C.

3. Presence of light

Light is a source of energy just like heat and can influence the rates of some chemical reactions by energising some of the molecules involved. For example, if chlorine and hydrogen are mixed in a glass container and the container placed in darkness, the rate of the reaction is almost negligible. When the container is placed in direct sunlight, the reaction rate speeds up so dramatically that the reaction mixture explodes.

Light is also vital to the photosynthetic production of starch by plants from carbon dioxide and water. *Photochemistry* is a branch of chemical kinetics that studies reactions influenced by light.

4. Surface area of the reactants

In the gas phase, the reactants are present as individual atoms or molecules. This state of subdivision offers the maximum opportunity for reaction. When one or more reactants are present in the solid phase, particle size influences the rate of the reaction. For example, aluminium foil reacts moderately with dilute sodium hydroxide solution when warmed but powdered aluminium reacts rapidly with cold sodium hydroxide solution and usually froths out of the test tube spontaneously. This happens because, for a given mass of metal, powder offers a much greater area to the reacting liquid than does foil.

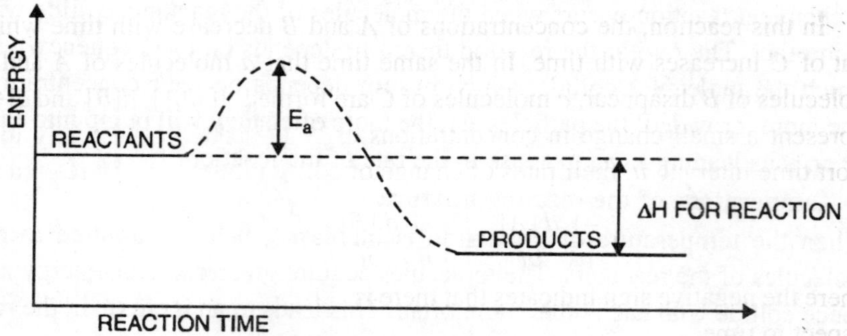

Figure 5.2 (a): The activation energy of an exothermic reaction.

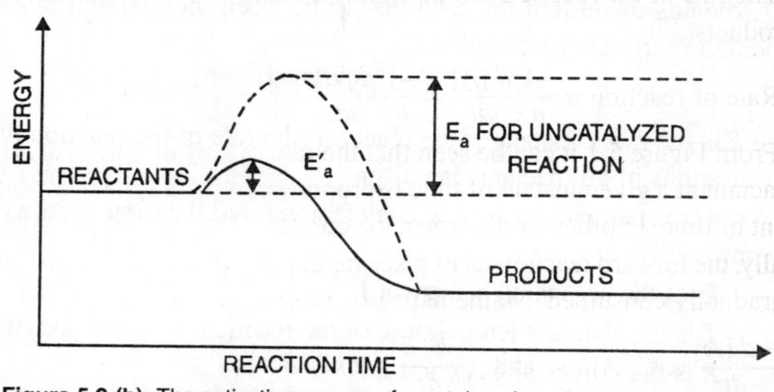

Figure 5.2 (b): The activation energy of a catalysed exothermic reaction.

5. Presence of a catalyst

A *catalyst* is a substance that alters the rate of a chemical reaction but remains *chemically* unchanged at the end of the reaction. Catalysts are often added to chemical reactions in small quantities to speed up their rates. Such catalysts are called *positive* catalysts. For example, the reaction between nitrogen and hydrogen to form ammonia is very slow under normal conditions. Finely divided iron is used along with high temperature and pressure to speed up the rate of the reaction in an industrial process known as Haber's process. Positive catalysts are able to affect the rates of reactions because they provide a new pathway for the reaction that has much lower activation energy E_a', as shown in Figure 5.2(b).

Negative catalysts or *inhibitors* retard the rate of a reaction. They are generally used to preserve certain substances. For example, the addition of a small quantity of anti-freeze like glycerol retards the rusting of iron machinery. Negative catalysts function by destroying the active intermediate products formed in a reaction, thus preventing the conversion of reactants into products.

5.3 Terminology used in the study of reaction kinetics

Consider the reaction

$$aA + bB \rightarrow cC \tag{5.2}$$

In this reaction, the concentrations of *A* and *B* decrease with time while that of *C* increases with time. In the same time that *a* molecules of *A* and *b* molecules of *B* disappear, *c* molecules of *C* are formed. If $d[A]$, $d[B]$ and $d[C]$ represent a small change in concentrations of *A, B* and *C* respectively in a short time interval *dt*, their rates of change of concentration will be related as

$$\frac{-1}{a}\frac{d\,[A]}{dt} = \frac{-1}{b}\frac{d\,[B]}{dt} = \frac{1}{c}\frac{d\,[C]}{dt} \tag{5.3}$$

where the negative sign indicates that there is a decrease in concentration with respect to time.

The rate of the reaction can be expressed in terms of any of the reactants or products.

$$\text{Rate of reaction} = \frac{-1}{a}\frac{d\,[A]}{dt} = \frac{-1}{b}\frac{d\,[B]}{dt} = \frac{1}{c}\frac{d\,[C]}{dt} \tag{5.4}$$

From Figure 5.1 it can be seen that the rate of concentration of a product or reactant at a given instant of time is proportional to its concentration at that instant of time. In other words, when the concentration of the reactants is high initially, the forward reaction takes place rapidly but slows down as the reactants are gradually consumed. Mathematically, this can be expressed as

$$\frac{-d\,[A]}{dt} \propto k'\,[A]^x \quad \text{and} \quad \frac{-d\,[B]}{dt} \propto k''\,[B]^y \tag{5.5}$$

where *x* and *y* are known as the *order* of the reaction with respect to *A* and *B*.

The numbers *x* and *y* denote how the rate of the reaction changes when the concentrations of the reactants are changed. For example, if $x = 1$ then the rate of the reaction is directly proportional to the concentration of *A*. In other words, doubling the concentration of *A* will also double the rate of the reaction. If $x = 2$, the rate of the reaction is proportional to the square of the concentration of *A*. In other words, doubling the concentration of *A* will increase the reaction rate by four times.

The overall rate of the reaction can be expressed as

$$\text{Rate} \propto [A]^x[B]^y$$

or

$$\text{Rate} = k\,[A]^x[B]^y \tag{5.6}$$

The overall order of the reaction is the sum of the orders of the reaction with respect to individual reactants, i.e. $(x + y)$.

Equation (5.6) is known as the *rate law* or the *rate equation.*

The proportionality constant k is known as the *rate constant* of the reaction.

The rate constant of a reaction can be defined as the rate of the reaction when the concentrations of all reactants are unity.

Molecularity of a reaction is the minimum number of molecules of the same or different reactants involved in the reaction. In the case of the reaction in equation (5.2) the molecularity is $(a + b)$.

The rate law of a reaction can only be determined experimentally. Hence the order and molecularity of a reaction may not always be equal.

The order of a reaction need not be a positive integer. It can be fractional, zero, or negative. For example, consider the reaction between hydrogen and bromine to give hydrogen bromide.

$$H_2 + Br_2 \rightarrow 2HBr$$

The rate law for this reaction is

$$\text{Rate} = \frac{k[H_2][Br_2]^{3/2}}{[Br_2] + k'[HBr]}$$

The order of the reaction may have non-integer values because the reaction may not proceed as given by the equation in a single step. The reaction may be composed of many elementary steps, which are of varying speeds. Each step in the reaction is called an *elementary step.* The slowest elementary step is called the *rate-determining step* of the reaction.

5.4 First Order Reactions

In a first order reaction, the rate of the reaction is directly proportional to the concentration of the reactant.

Consider the reaction

$$A \rightarrow B + C \tag{5.7}$$

Let a mol dm^{-3} be the initial concentration of A. After a certain time t, x mol dm^{-3} of A decomposes to form B and C. The rate of formation of B and C is thus dx/dt, which is proportional to the instantaneous concentration of A.

$$dx/dt = k(a - x) \tag{5.8}$$

Integrating the equation, we get

$$kt = \ln a/(a - x) \tag{5.9}$$

$$k = \frac{2.303}{t} \log \frac{a}{(a - x)} \tag{5.10}$$

In equation (5.10) the factor $\log a/(a - x)$ is a number and therefore the units of the rate constant are time^{-1}. The value of the rate constant is independent of the initial concentration of the reactant.

5.4.1 Half-change time

An important test for a first order reaction is that the time required to halve the concentration of the reactant should be constant and independent of the initial concentration of the reactant.

If we substitute $x = a/2$ at time $t = t_{1/2}$ in equation (5.10), we get

$$k = \frac{2.303}{t_{1/2}} \log \frac{a}{(a - a/2)}$$

$$= 2.303 \log 2$$

$$k = \frac{0.693}{t_{1/2}} \quad (5.11)$$

$t_{1/2}$ is the time required for the concentration of the reactant to become half and is therefore known as half-change time. In the case of radioisotopes, the half-change time, i.e., the time required for an isotope to decay to half its original quantity, is a characteristic property of the isotope and is called *half-life* of the isotope.

5.4.2 Pseudo-unimolecular reactions

Consider the hydrolysis of methyl acetate in aqueous solution.

$$CH_3COOCH_3 + H_2O \rightarrow CH_3OH + CH_3COOH$$

The reaction rate is too slow to be observed under normal conditions. It can be catalysed by the addition of H^+ ions from a mineral acid such as hydrochloric acid. In this reaction water is present in such a large excess that its concentration does not change appreciably over the course of the reaction. Hence the reaction rate is solely dependant on the concentration of methyl acetate. The reaction though bimolecular is of first order. Such reactions are called *pseudo-unimolecular* reactions.

During the course of this reaction, the acidity of the reaction mixture increases due to the formation of acetic acid. Thus the reaction can be studied by dissolving methyl acetate in water containing a small amount of dilute hydrochloric acid. A definite volume of the reaction mixture is withdrawn at regular intervals and titrated against standard alkali. The volume of alkali solution required for titration will be equivalent to the amount of hydrochloric acid present and the acetic acid produced in the reaction.

At the commencement of the reaction at time $t = 0$, the volume of alkali V_0 will be equivalent to the amount of HCl alone. Hence if the volume of the alkali at time $t = V_t$ then $(V_t - V_0)$ will be a measure of the amount of acetic acid produced in the reaction and hence of the methyl acetate decomposed so far. If V_∞ is the volume of the alkali when the reaction is complete then $(V_\infty - V_0)$ gives the initial concentration of methyl acetate.

Thus the rate constant of the reaction is given by equation (5.10) where $a \propto (V_\infty - V_0)$ and $x \propto (V_t - V_0)$.

Therefore

$$k = \frac{2.303}{t} \log \frac{(V_\infty - V_0)}{(V_\infty - V_t)} \quad (5.12)$$

5.5 Second order reactions

In a second order reaction, the rate of a reaction is directly proportional to the concentration of two reactants.

Consider the reaction

$$A + B \,\square\, C + D \quad (5.13)$$

The rate law for this reaction can be written as

$$\text{Rate} \propto [A][B] \quad (5.14)$$

Let the initial concentrations of A and B be a mol dm^{-3} and b mol dm^{-3} respectively. If after time t, x mol dm^{-3} of C and D are formed then the rate law can be written as

$$dx/dt = k(a-x)(b-x) \quad (5.15)$$

where k is the second order rate constant.

Integrating this equation we get

$$k = \frac{1}{t(a-b)} \ln \frac{b(a-x)}{a(b-x)} \quad (5.16a)$$

or

$$k = \frac{2.303}{t(a-b)} \ln \frac{b(a-x)}{a(b-x)} \quad (5.16b)$$

If the initial concentrations of A and B are equimolar, the rate law can be expressed as

$$dx/dt = k(a-x)(a-x) = k(a-x)^2 \quad (5.17)$$

This rate law is also applicable when the rate of the reaction is proportional to the square of the concentration of one substance, as for example in the following reaction:

$$2A \rightarrow C + D \quad (5.18)$$

Integrating eqn. (5.17), we get

$$k = \frac{x}{ta(a-x)} \quad (5.19)$$

From equations (5.15a) and (5.19) it can be seen that the units of the rate constant are concentration^{-1} time^{-1}.

5.5.1 Half-change time

Substituting $x = a/2$ when $t = t_{1/2}$ in eqn. (5.19), we get

$$k = \frac{1}{t_{1/2}} = \frac{a/2}{a(a - a/2)}$$

$$= \frac{1}{t_{1/2}a}$$

$$t_{1/2} = \frac{1}{ka} \tag{5.20}$$

Equation (5.20) shows that the time for half change varies inversely as the initial concentration of the reactant.

The hydrolysis of an ester by an alkali is an example of a second order reaction. (This is in contrast to the hydrolysis of an ester by an acid, which is a first order, pseudo-unimolecular reaction (Section 5.4.2).) The alkali hydrolysis of higher fatty acids like stearic, palmitic and oleic acids is used industrially for the manufacture of soap and is called *saponification.*

Consider the hydrolysis of ethyl acetate by dilute sodium hydroxide solution. This reaction can be studied by mixing equimolar quantities of sodium hydroxide and dilute ethyl acetate solution that are at the same temperature. The reaction mixture is kept in a thermostat and a conductivity cell is dipped in it. The conductivity of the solution is determined at regular intervals. The initial conductivity reading G_0 will be a measure of the sodium hydroxide initially present since it is the only conducting species present in the solution. As the reaction proceeds, the conductivity G_t of the reaction mixture at time t rises due to the formation of the salt sodium acetate. Hence the quantity $(G_t - G_0)$ is a measure of the extent of hydrolysis of the ester at time t. If G_∞ is the conductivity of the reaction mixture when the reaction is complete then $(G_\infty - G_0)$ is the initial concentration of ethyl acetate.

Thus the rate constant of the reaction can be determined by substituting in equation (5.19).

$$k = \frac{(G_1 - G_0)}{t(G_\infty - G_0)(G_\infty - G_t)} \tag{5.21}$$

5.6 Reactions of nth Order

Consider a general reaction

$$nA \rightarrow \text{Products} \tag{5.22}$$

The rate law for an nth order reaction can be written as

$$dx/dt = k(a - x)^n \tag{5.23}$$

where a is the initial concentration of A

x is the concentration of A at time t

n is the overall order of the reaction.

Integrating this equation we get

$$kt = \frac{1}{(n-1)}\left[\frac{1}{(a-x)^{n-1}} - \frac{1}{a^{n-1}}\right] \quad (5.24)$$

where $n \neq 1$.

The units of rate constant are concentration$^{(1-n)}$ time $^{-1}$.

5.7 Determination of the Order of a Reaction

The order of a reaction can only be determined experimentally. It cannot be derived directly from the stoichiometry of the reaction equation in most cases. There are several methods available for determining the order of a reaction. Some of these are discussed below.

5.7.1 Half-change time method

This method is only applicable for reactions whose overall order is not unity because for first-order reactions, the half-change time is independent of the initial concentration. For all other orders of reaction, the half change time $t_{1/2}$ is given by

$$t_{1/2} = \frac{2^{(n-1)} - 1}{k^{n-1}} \cdot \frac{1}{a^{n-1}} \quad (5.25)$$

where a is the initial concentration of the reactants and n is the order of the reaction.

Taking logarithms on both sides,

$$\log t_{1/2} = \log \frac{2^{(n-1)} - 1}{k^{n-1}} + (1-n) \log a \quad (5.26)$$

A plot of log $t_{1/2}$ versus log a gives a straight line with slope $(1 - n)$. To use this method, a graph of concentration of the reactant versus time is plotted, as shown in Figure 5.3(a). On this graph, a point a' is chosen and the point where the concentration has become $a'/2$ is found. The time interval between these two points is the half-change time $t_{1/2}$ for the initial concentration a'. This process is repeated a number of times for different initial concentrations. A graph of log $t_{1/2}$ versus log of the corresponding initial concentration is plotted, as shown in Figure 5.3(b). The slope of this graph is $(1 - n)$.

This method can give better results if the reaction is followed to a higher degree of completion. For every value of initial concentration a', the time taken for it to fall to 0.75 a' is determined. This time interval is called fractional life t_α. A plot of log t_α versus log a' will give a straight line of slope $(1 - n)$.

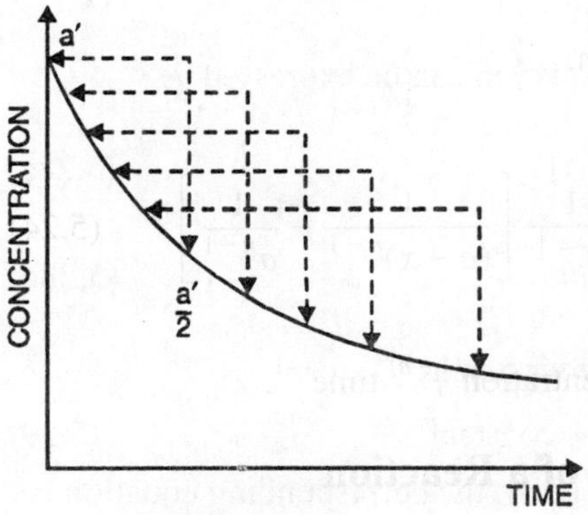

Figure 5.3(a)

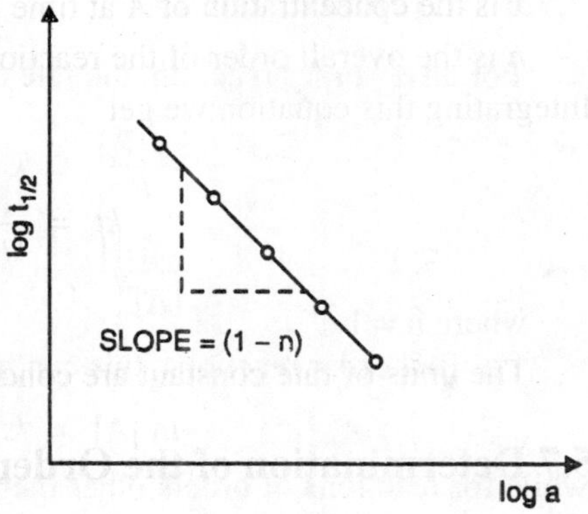

Figure 5.3(b)

5.7.2 *Initial-rate method*

In this method, the initial rate of the reaction R_0 is measured for several runs. In each run the initial concentration of one reactant is varied such that by making several runs, the initial concentration varies over a wide range of values. If the rate law has the form of equation (5.6) then for an initial rate R_0

$$\log R_0 = \log k + x \log [A]_0 + y \log [B]_0 \tag{5.27}$$

A plot of $\log R_0$ versus $\log [A]_0$ at constant $[B]_0$ has slope x. The order of the reaction with respect to reactant B can be found similarly.

5.7.3 *Ostwald's Isolation method*

In this method, all the reactants excepting one (say, A) are taken in excess. The order of the reaction then corresponds only to the order with respect to A, since the initial concentrations of the other reactants are so high that they do not change appreciably during the course of the reaction. This process is repeated for each reactant and the order of the reaction with respect to that reactant is determined. The overall order of the reaction will be the sum of all these orders.

5.7.4 *Substitution in the integrated rate laws*

In this method, the values of the initial concentration a and the concentrations after time t are determined for a number of values of time t. These values are substituted into all the different rate equations. The rate equation for which a constant value of the rate constant is obtained gives the order of the reaction. This method involves trial-and-error and is quite inaccurate. It should only be used to confirm the order of the reaction that has already been obtained by some other method.

5.7.5 Graphical method

For an nth order reaction, the rate equation can be expressed as

$$\frac{-d[A]}{dt} = k\,[A]^n \tag{5.28a}$$

or

$$\frac{-d[A]}{[A]^N} = k\,dt \tag{5.28b}$$

For first order reactions, this can be written as

$$-\ln [A] = kt + \text{constant} \tag{5.29}$$

while for reactions of orders other than unity, the corresponding equation is

$$\frac{1}{[A]^{(n-1)}} = kt + \text{constant} \tag{5.30}$$

A plot of $-\ln [A]$ versus t gives a straight line if the reaction is of first order. A plot of $1/[A]^{(n-1)}$ versus t will be, a straight line if the order of the reaction is other than unity.

The graphical method involves plotting all these curves and determining which is the most linear. This is a trial-and-error method and can be erroneous because it is often very hard to decide which of the plots is most nearly linear, especially if the reaction has not been followed very far.

As an example of a reaction whose order has to be determined, consider the reaction between acetone and iodine, which can be represented as

$$CH_3COCH_3 + I_2 \rightarrow CH_3COCH_2I + H^+ + I^-$$

The overall reaction is first order in acetone and acid but is independent of the concentration of iodine. This can be established by preparing a reaction mixture containing 20 cm^3 of 2 M acetone, 10 cm^3 of 1 M sulphuric acid, 20 cm^3 of 0.05 M I_2 in 10% KI solution and 150 cm^3 water. Aliquots of the reaction mixture are removed at fixed time intervals. The reaction is stopped by adding sodium acetate solution. The iodine in the aliquot is titrated with standard sodium thiosulphate solution using starch as indicator. The burette reading is a measure of how far the reaction had progressed at the time of the withdrawal of the aliquot. Another reaction mixture similar to the first but containing only half the amount of iodine is prepared and the reaction is followed similarly. A graph of burette reading versus time is plotted for both the runs. The slopes of the straight lines obtained are determined. The order, α, of the reaction with respect to iodine can be determined by the equation

$$\log (S_1 / S_2) = \alpha \log ([Ac]_1 / [Ac]_2)$$

where S_1 and S_2 are the slopes, and $[Ac]_1$ and $[Ac]_2$ are the concentrations of acetone in the two runs.

The orders with respect to sulphuric acid and iodine can be determined similarly.

5.8 Determination of the Rate Constant of the Reaction

Once the order of the reaction with respect each reactant has been determined, the rate constant of the reaction can be determined from the slope of an appropriate graph. For example, if the reaction is of first order than equation (5.10) shows that

$$\log \{ a/(a-x) \} = kt / 2.303 \tag{5.31}$$

A graph of $\log\{a/(a-x)\}$ versus t would give a straight line of slope $k/2.303$.

Another method is to substitute the values of the initial concentration a and the subsequent concentrations x for each observation at time t to get a value of k for every observation. These k values are averaged to yield a mean value of the rate constant. This method is generally inaccurate.

5.9 Reactions in Solution

Reactions in solution differ from reactions in the gas phase by the presence of a solvent. The classical kinetic theory discussed so far does not take into account the effect of a solvent on the rate of a reaction. It is possible for the solvent to strongly affect the reaction rate through one or more of its following properties.

1. Degree of Solvation

Reactants in solution are generally surrounded by solvent molecules. This phenomenon is called solvation. It may prevent the reacting molecules from encountering each other, thus slowing down the reaction. The more the number of solvent molecules surrounding each reactant molecule, the higher is the degree of solvation. Different solvents give rise to different degrees of solvation, thus resulting in differing values of the rate constant.

2. Dielectric constant

If the reactants are ionic or polar molecules, the electrostatic forces between them will depend on the dielectric constant of the solvent. The higher the dielectric constant of the solvent, the faster is the reaction between ionic and polar reactants.

3. Viscosity

Reactions in solution are very fast. They are generally over in about a millionth of a second. However, the reaction rate can be limited by the rate at which two reactant molecules can diffuse through the solvent to encounter each other. The greater the viscosity of the solvent, the slower is the reaction.

4. Hydrogen boding

A hydrogen bond is a weak bond of the form $A - H \cdots B$ where A and B are highly electronegative elements and B possesses a lone pair of electrons. The hydrogen bonding that can occur between a solvent and a reactant can affect the reaction rate.

Reactant molecules diffuse towards each other slowly through the solvent and then linger near each other for a while. The reactant molecules are surrounded by a cage of solvent molecules that keep them close together. During the relatively long time that the reactant molecules are close to each other, they collide with each other repeatedly. An encounter pair may form, which is a loosely bound complex of the reacting species. This encounter pair may go on to form the products or it may break up without reaction.

If the rate at which the encounter pair breaks up is much less than the rate at which it forms products then the reaction is said to be *diffusion-controlled.* The rate of a diffusion-controlled reaction is governed by the rate at which the reactant molecules diffuse throughout the solvent.

If the rate at which the encounter pair forms products is much less than the rate at which it breaks up then the reaction is said to be *activation-controlled.* In an activation-controlled reaction, the reaction proceeds at the rate at which the encounter pair gathers activation energy from the surrounding solvent.

5.9.1 The kinetic salt effect

For ionic reactions in solution, the encounter pair formed is a highly charged ionic complex.

Consider the reaction

$$A + B \longleftrightarrow [\,C^{\ddagger}\,] \rightarrow \text{products} \tag{5.32}$$

$$\text{Rate of reaction} = k^{\ddagger}\, C^{\ddagger} = k_2\,[A]\,[B] \tag{5.33}$$

where k_2 is the second order rate constant of the reaction

$$A + B \longleftrightarrow [\,C^{\ddagger}\,]$$

and $k^{\ddagger}$ is the second order rate constant of the reaction

$$[\,C^{\ddagger}\,] \rightarrow \text{products}$$

Ionic solutions are highly non-ideal, even at low concentrations, due to the various effects exerted by the solvent, as described in Section 5.9. Hence the thermodynamic equilibrium constant of the above reaction is expressed in terms of the activities of the ionic species involved.

$$K = \frac{a_c^{\ddagger}}{a_A a_B} \tag{5.34}$$

The activities are further expressed in terms of activity coefficients and actual concentrations as

$$K = \frac{\gamma_C^{\ddagger}\,[C^{\ddagger}]}{\gamma_A[A]\,.\,\gamma_B[B]}$$

$$K = \frac{K_{\gamma}[C^{\ddagger}]}{[A][B]} \tag{5.35}$$

where
$$K_\gamma = \gamma_C{}^{\ddagger}/\gamma_A \cdot \gamma_B \tag{5.36}$$

$$[C^{\ddagger}] = \frac{K}{K_\gamma}[A][B] \tag{5.37}$$

Substituting equation (5.37) in equation (5.33),

$$\frac{k^{\ddagger} K\,[A]\,[B]}{K_\gamma} = k_2\,[A]\,[B]$$

$$k_2 = k^{\ddagger} K / K_\gamma \tag{5.38}$$

When the activity coefficients are unity, i.e., $K_\gamma = 1$, the ideal second order rate constant k_2° is obtained.

$$k_2 = k_2^\circ K \tag{5.39}$$

Substituting equation (5.39) in equation (5.38),

$$k_2 = k_2^\circ / K_\gamma \tag{5.40}$$

Taking logarithms on both sides of equation (5.36), we get

$$\log K_\gamma = \log \gamma_C{}^{\ddagger} - \log \gamma_A - \log \gamma_B \tag{5.41}$$

At low concentrations, the activity coefficients can be expressed in terms of the ionic strength I of the solution using the Debye-Huckel Limiting Law (equation (3.04)). If the ionic charges on ions A and B are z_A and z_B then the ionic charge on the complex is $(z_A + z_B)$. Expressing the activity in equation (5.41) in terms of the Debye-Huckel Limiting Law,

$$\log K_\gamma = -A\,\{\,(z_A + z_B)^2 - z_A{}^2 - z_B{}^2\,\}\sqrt{I} \tag{5.42}$$

$$\log K_\gamma = -2Az_Az_B\sqrt{I} \tag{5.43}$$

Substituting equation (5.43) in equation (5.40),

$$\log k_2 = \log k_2^\circ - \log K_\gamma$$

$$\log k_2 = \log k_2^\circ + 2Az_Az_B\sqrt{I} \tag{5.44}$$

Equation (5.41) shows the variation of the rate constant of an ionic reaction in solution with the ionic strength of the solution. If the reactant ions have the same charge sign then z_Az_B is positive and the rate constant increases with increasing ionic strength. On the other hand, if the reactant ions have opposite charge signs then z_Az_B is negative and the rate constant decreases with decreasing ionic strength. This phenomenon is called the *kinetic salt effect.*

5.10 Autocatalysis

Catalysis is the phenomenon in which the rate of a chemical reaction is altered by the presence of a catalyst. Some chemical reactions are catalysed by one of the products of the reaction itself. Such reactions are called auto-catalysed reactions and the phenomenon is called autocatalysis.

A spectacular example of autocatalysis is the hydrogen bomb. A neutron collides with a uranium nucleus to release a tremendous amount of energy and two or three neutrons. These neutrons collide with other uranium atoms to release two or three neutrons *each*, thus triggering off an explosive chain reaction.

The decolourisation of potassium permanganate by oxalic acid in acidic medium is another example of an autocatalytic reaction.

$$2MnO_4^- + 16H^+ + 5C_2O_4^- \rightarrow 2Mn^{2+} + 10CO_2 + 8H_2O$$

The Mn^{2+} ions are generated in the reaction act as catalyst. The decolourisation is slow in the beginning. As the concentration of Mn^{2+} ions increases with time, the reaction accelerates and the solution decolourises rapidly. This behaviour is shown in Curve *I* of Figure 5.4.

When Mn^{2+} ions are included in the initial reaction mixture, the reaction accelerates right from the start, as shown in Curve *II* of Figure 5.4.

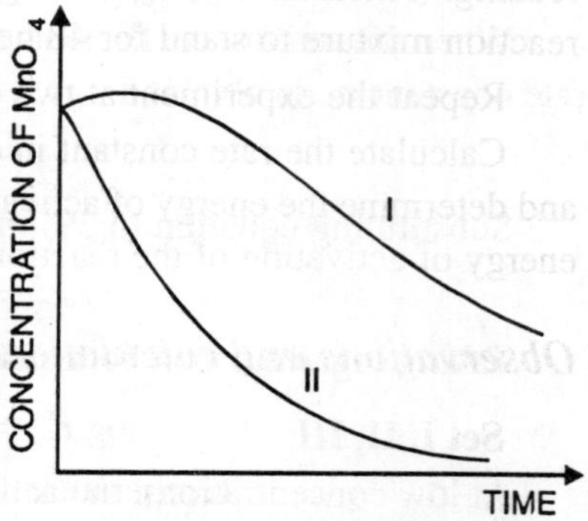

Figure 5.4: Graph of concentration of MnO_4^- ions against time

The reaction can be studied by removing aliquots of the thermostated reaction mixture at fixed intervals of time. The reaction is arrested by adding potassium iodide to the aliquot. The iodine released is titrated with sodium thiosulphate solution, using starch as indicator. The burette reading serves as a measure of the MnO_4^- present in the reaction mixture at the time the reaction was stopped.

EXPERIMENTS

Experiment 5.1

Aim: To determine the energy of activation of hydrolysis of methyl acetate catalysed by hydrochloric acid.

Requirements

Methyl acetate, 0.5N HCl, 0.1N NaOH, thermostat, pipettes, burettes etc.

Procedure

Take 50 cm^3 of 0.5N HCl in a dry 250 cm^3 comical flask clamped in a thermostat bath at 30 °C and cork it. Also place a stoppered tube containing 10 cm^3 of methyl acetate in the same bath. When the temperature equilibrium is reached, pipette out exactly 5 cm^3 of methyl acetate into the flask containing acid and note the time. Shake thoroughly and immediately withdraw 5 cm^3 of the sample of the mixture and run into conical flask containing about 25 cm^3 of ice-cold water to arrest the reaction. Titrate as early as possible with 0.1N NaOH. Repeat titration of further 5 cm^3 samples at intervals of 5, 10, 15, 20, 30, 40 and 50 minutes from the start.

Heat the remaining mixture in a conical flask on a water bath at 100°C for about 1 hour to ensure complete hydrolysis. Cool the flask to room temperature and titrate 5 cm^3 of the mixture with 0.1N NaOH solution. This gives T_∞ reading. (Alternatively, T_∞ reading can be taken after allowing the remaining reaction mixture to stand for 48 hours at room temperature.)

Repeat the experiment at two different temperatures: 35°C and 40°C.

Calculate the rate constant in each case. Plot a graph of log k versus T^{-1} and determine the energy of activation ΔE from the slope. Hence evaluate the energy of activation of the reaction.

Observations and calculations

Set I, II, III

Temperature = ____°C = ____ K

T_0 = ____cm^3 T_∞ = ____cm^3

$a = T_\infty - T_0$ = ________ cm^3

Time t (min)	Vol. of NaOH (cm^3)	x = $(T_t - T_0)$ (cm^3)	$a - x$ (cm^3)	log $(a/a - x)$	K = $(2.303/t)\log(a/a - x)$ (min^{-1})

Mean K = ________min^{-1}

From the graph of log $(a - x)$ versus t,
for Set I, II, III $K = 2.303 \times$ slope = ________ (min^{-1})

Temperature T / (K)	K /(min^{-1})	log K	1/T/ (K^{-1})

From the graph of log K versus T^{-1}, slope = ________

$\Delta E = — 2.303 \times \text{slope} \times R$ = ________ kJ mol^{-1}

log A = log $K + (\Delta E/2.303RT)$	ΔS = $2.303R\{\log A - \log(RT/Nh)$ (J mol^{-1} K^{-1})	$\Delta G = \Delta E - T\Delta S$ (kJ mol^{-1})

Result

Temp (*K*)	*K* by calculation (min^{-1})	*K* by graph (min^{-1})	ΔE (kJ mol^{-1})	log *A*	ΔS (J)	ΔG (kJ mol^{-1})

Experiment 5.2

Aim: To study the influence of ionic strength on reaction between potassium persulphate and potassium iodide solution.

Requirements

0.01M $K_2S_2O_8$, 0.4M KI, 0.4M KCl, standard 0.001N $Na_2S_2O_3$, starch solution.

Procedure

Dilute 25 cm^3 of KI solution to 100cm^3 in standard flask and place in a thermostat. Take 25 cm^3 $K_2S_2O_8$ solution in a 250 cm^3 conical flask and place it in the thermostat.

Add the potassium iodide solution to potassium persulphate solution and record the time. Mix well. Pipette out 10 cm^3 of mixture and freeze reaction by adding ice and water. Titrate against standard sodium thiosulphate solution. Take readings at time intervals of 2, 5, 10, 15, 20, 25 and 30 minutes from the start.

Repeat the set with addition of potassium chloride solution to alter the ionic strength. To 25 cm^3 of potassium iodide solution, add 25, 50 and 75 cm^3 of potassium chloride solution and sufficient volume of water to make 100 cm^3, if necessary.

For different sets, plot *t* versus log $(a - x)$ and evaluate the rate constant, *k*. The terms *a* and *x* may be expressed in terms of concentration of $Na_2S_2O_3$ solution, *a* being calculated from initial concentration of potassium thiosulphate. Determine the slope and hence *k* and test validity of equation

$$\log k = \log k^{\circ} + 2\sqrt{I}.$$

Observations and calculations

Calculation of *a*

Initial concentration of $K_2S_2O_8$ = 0.01 × 25/125 = 0.002M = 0.004N

$$S_2O_8^{2-} + 2I^- \rightarrow I_2 + 2SO_4^{2-}$$

0.004N $S_2O_8^{2-}$ will give 0.004N I_2

$$I_2 + 2S_2O_8^{2-} \rightarrow 2I^- + S_4O_6^{2-}$$

Volume of 0.001N $Na_2S_2O_3$ required to titrate 10 cm^3 of 0.004N I_2 solution =

$$a = 0.004 \times 10/0.001 = 40 \text{ cm}^3$$

Titration

Solution in burette: 0.001N (standard) $Na_2S_2O_3$

Solution in conical flask: 10 cm^3 of reaction mixture

Indicator used: starch solution

End point: blue to colourless

Set I, II, III, IV

Time t/(min)	Vol of 0.001N $Na_2S_2O_3$/(cm^3)	$a-x$ /(cm^3)	log ($a/a-x$)	k = (2.303/t)log($a/a-x$) /(min^{-1})
2				
5				
10				
...				

Mean k = ________ min^{-1}

Set no.	Final Concentration (M)			$I = \frac{1}{2}\Sigma c_i z_i^2$ (M)	$\sqrt{I}$	log K
	$K_2S_2O_8$	KI	KCl			

From the graph of log ($a/a-x$) versus time t,

for Set I, II, III, IV $K = 2.303 \times$ slope = ________ (min^{-1})

From the graph of log K versus $\sqrt{I}$, slope = ________

Result

The observed results show that the rate constant of the reaction increases with increase in ionic strength.

The slope of the graph of log K versus $\sqrt{I}$ is ________.

For the equation $\log K = \log K_a + 2\sqrt{I}$ to be valid, the slope should be 2. This expected value is not obtained because according to the Debye-Huckel theory the equation holds good for only very dilute solutions ($\approx 10^{-4}$ M), which are not used in this experiment. Also, the ice used to quench the reaction mixture should be made with distilled water otherwise it increases the ionic strength of the solution as it melts due to which dissolved salts go into the solution.

Experiment 5.3

Aim: Investigation of the reaction between acetone and iodine.

Requirements

2 M acetone, 1 M sulphuric acid, 0.05 M iodine in 10% potassium iodide solution, 1 M sodium acetate solution, standard 0.01 M sodium thiosulphate solution.

Procedure

Prepare the following mixtures at room temperature from solutions kept in a large water base at room temperature:

Set no.	Acetone	Water	Sulphuric acid	I_2 in 10% KI solution
I	20 cm^3 of 2 M soln	150 cm^3	10 cm^3 of 1 M soln	20 cm^3 of 0.05 M soln
II	20 cm^3 of 1 M soln	150 cm^3	10 cm^3 of 1 M soln	20 cm^3 of 0.05 M soln
III	20 cm^3 of 2 M soln	150 cm^3	10 cm^3 of 0.5 M soln	20 cm^3 of 0.05 M soln
IV	20 cm^3 of 2 M soln	150 cm^3	10 cm^3 of 1 M soln	20 cm^3 of 0.025 M soln

Remove 20 cm^3 of the sample from the reaction mixture at 5 minutes interval and run into 10 cm^3 of sodium acetate solution. Titrate the residual iodine against standard 0.01 M sodium thiosulphate using starch indicator.

Keeping the total volume constant, investigate the effect of changes in concentration of iodine, acetone and sulphuric acid on the rate of the reaction. Plot a graph of titre volume of sodium thiosulphate (equivalent to the residual iodine) against time for each mixture. Determine the order of the reaction with respect to acetone, iodine and sulphuric acid.

Observations and calculations

Solution in burette: standard 0.01 M sodium thiosulphate solution

Solution in conical flask: I_2 solution + 10 cm^3 1 M sodium acetate solution

Indicator used: starch solution

End point: blue to colourless

Time *t*	Burette reading (cm^3)			
(min)	Set I	Set II	Set III	Set IV
5				
10				
15				
20				
25				
30				

From the graphs of volume of 0.01 M sodium thiosulphate added versus time,

Slope of graph for Set I = S_1 = ______

Slope of graph for Set II = S_2 = ______

Slope of graph for Set III = S_3 = ______

Slope of graph for Set IV = S_4 = ______

In Set I, concentration of acetone = $[Ac]_1$ = ______ M

In Set II, concentration of acetone = $[Ac]_2$ = ______ M

In Set I, concentration of hydrogen ions = $[H^+]_1$ = ______ M

In Set III, concentration of hydrogen ions = $[H^+]_2$ = ______ M

In Set I, concentration of iodine = $[I_2]_1$ = ______ M

In Set IV, concentration of iodine = $[I_2]_2$ = ______ M

Order of the reaction with respect to acetone

$= \{\log S_1 - \log S_2\}/\{\log [Ac]_1 - \log [Ac]_2\}$

= ______

Order of the reaction with respect to acid

$= \{\log S_1 - \log S_3\}/\{\log [H^+]_1 - \log [H^+]_2\}$

= ______

Order of the reaction with respect to iodine

$= \{\log S_1 - \log S_4\}/\{\log [I_2]_1 - \log [I_2]_2\}$

= ______

Result

Order of the reaction between acetone and iodine with respect to

(*i*) acetone = α = ______

(*ii*) sulphuric acid = β = ______

(*iii*) iodine = γ = ______

$$\text{Rate} = k\,[\text{acetone}]^{\alpha}\,[\text{sulphuric acid}]^{\beta}\,[\text{iodine}]^{\gamma}$$

Experiment 5.4

Aim: To determine conductometrically the second order rate constant for the hydrolysis of ethyl acetate by sodium hydroxide.

Requirements

Conductivity meter, dip-type conductometric cell, 0.05 M NaOH, 0.2 M ethyl acetate, 0.01 M sodium acetate (prepared by neutralising 20 cm^3 of 0.05 M NaOH solution with acetic acid and making the volume up to 100 cm^3. All the solutions should be made using distilled water free of carbon dioxide.

Procedure

Measure the conductivity of 0.01 M NaOH and 0.01 M sodium acetate solutions at room temperature. These represent the initial conductance G_0 and final conductance G_∞ of the mixture respectively. In a beaker take 20 cm^3 of NaOH solution and 50 cm^3 distilled water. In another beaker take 5 cm^3 of ethyl acetate and 25 cm^3 distilled water. Place the beakers in a thermostat and allow temperature equilibrium to be reached. Mix the two solutions, shake thoroughly and place the conductivity cell in the mixture. Take conductance readings at one-minute intervals for 30 minutes. Plot a graph of $(G_0 - G_t)/(G_\infty - G_t)$ versus time *t* in minutes. Calculate the rate constant *k*.

Observations and calculations

$$G_0 = \_\_\_\_\_ S$$
$$G_\infty = \_\_\_\_\_ S$$

Time *t* (min)	G_t / (*S*)	$(G_0 - G_t)/(G_\infty - G_t)$
1		
2		
...		
30		

From the graph of $(G_0 - G_t)/(G_\infty - G_t)$ versus time, slope = _____
Slope = *k a*
where *a* is the initial concentration of ethyl acetate = _____ mol dm^{-3}
k = _____ $mol^{-1}\ dm^3\ s^{-1}$

Result

The rate constant of the reaction between sodium acetate and sodium hydroxide is ____ $mol^{-1}\ dm^3\ s^{-1}$.

Experiment 5.5

Aim: To investigate the autocatalytic reaction between potassium permanganate and oxalic acid.

Theory

The reaction between potassium permanganate and oxalic acid in presence of a mineral acid takes place according to the following equation:

$$2KMnO_4 + 5H_2C_2O_4 + 3H_2SO_4 \rightarrow K_2SO_4 + MnSO_4 + 10CO_2 + 8H_2O$$

The reaction is catalysed by manganese ions formed during the reaction. The progress of the reaction can be followed by estimating the residual permanganate concentration from time to time. This is done by running the samples of the mixture to excess of KI and then titrating the liberated iodine with standard $Na_2S_2O_3$ solution. A plot of titre value of sodium thiosulphate against time reveals the autocatalytic effect of Mn^{2+} ions. The reaction in the beginning is very slow but the rate increases rapidly due to the formation of Mn^{2+} ions. If Mn^{2+} ions are added to the initial reaction mixture, the titre values fall rapidly from the very beginning.

The order of the reaction with respect to potassium permanganate can be determined from these curves by determining the time required for half change (half-life method). If the order of the reaction is n, then the half-life of the reaction $t_{1/2}$ depends on the initial concentration $[A]_0$ of the reactant according to the equation

$$\log t_{1/2} = \frac{\log 2^{n-1} - 1}{(n-1)k_A} - (n-1)\log [A]_0$$

A plot of $\log t_{1/2}$ versus $\log [A]_0$ gives a straight line of slope $(1 - n)$. The half-life method can be improved by the use of fractional life t_α which is defined as the time required for $[A]_0$ to fall to $\alpha[A]_0$. A convenient value of α is 0.75.

Requirements

Thermostat, 0.1M oxalic acid, 1 M sulphuric acid, 0.2M manganese (II) sulphate, 0.02M potassium permanganate, 0.01M sodium thiosulphate, 5% potassium iodide, starch solution.

Procedure

Prepare the following reaction mixtures and thermostat them in a water bath:

Reaction Mixtures	Volume of (cm^3)				
	0.1M oxalic acid	0.02 M $KMnO_4$	0.2M $MnSO_4$	1M H_2SO_4	Water
I	80	20	0	10	15
II	80	10	0	10	25
III	80	20	2	10	13
IV	80	20	2	10	138

At time t_0 add the required volume of $KMnO_4$ to the mixture and start the stopwatch. Withdraw 10 cm^3 of the reaction mixture at regular time intervals and run into 5 cm^3 of 5% KI taken in a conical flask. For reaction mixtures I and II, the time interval can be taken as 60 seconds, while for reaction mixtures III and IV, it can be taken as 30 seconds. Cover the mouth of the conical flask and keep it in the dark for 5 minutes till the iodine is fully liberated. Titrate the liberated iodine with standard sodium thiosulphate solution using starch solution as the indicator. The titre value of sodium thiosulphate is equivalent to the concentration of potassium permanganate in the reaction mixture at that time.

Plot the titre values of sodium thiosulphate against time for each mixture. Using the curve for reaction mixture III, determine the order of the reaction by the fractional life method. Choose an initial concentration $[A]_0'$ on the curve and find the point where $[A]_0'$ has fallen to $0.75[A]_0'$. The time interval between these two points is the fractional life $t_{0.75}$ for the initial concentration $[A]_0$. Then pick another point $[A]_0''$ and determine $t_{0.75}$ for this initial concentration. Repeat this process to get around five pairs of $[A]_0$ and $t_{0.75}$ values. Plot a graph of $\log t_{0.75}$ versus $\log [A]_0$. The slope will have value $(1 - n)$ where n is the order of the reaction.

Observations and calculations

Room temperature = ____ °C

Time	Titre reading (cm^3)	
(min)	Reaction Mixture I	Reaction Mixture II
1		
2		
...		
10		

Time	Titre reading (cm^3)	
(min)	Reaction Mixture III	Reaction Mixture IV
0.5		
1.0		
...		
5.5		

Determination of order of reaction by fractional life method.

Curve III

Initial conc $[A]_0$	Final conc. 0.75 $[A]_0$	Time interval $t_{0.75}$ (min)	$\log t_{0.75}$	$\log [A]_0$

From the graph of $\log t_{0.75}$ versus $\log [A]_0$, slope = ____

Slope = $1 - n$

n = ______

Result

The order of the reaction with respect to potassium permanganate is ______.

Experiment 5.6

Aim: **To investigate the reaction between potassium persulphate and potassium iodide by colorimetric measurements.**

Requirements

Colorimeter, saturated solution of potassium persulphate, 0.01M KI, 0.05M KI, 0.05M $K_2S_2O_8$.

Theory

The reaction between potassium iodide and potassium persulphate is of second order.

$$K_2S_2O_8 + 2KI \rightarrow 2K_2SO_4 + I_2$$

The concentration of iodine increases with time, which leads to an increase in the colour intensity of the solution with time. Hence the absorption of the solution also increases with time. The concentration of iodine liberated can be determined by colorimetric measurements with respect to time. Hence the rate constant of this second order reaction can be calculated by the formula:

$$K = \frac{1}{t} = \frac{x}{a(a-x)}$$

where x is the concentration of iodine liberated

a is the initial concentration of KI and $K_2S_2O_8$.

To calculate x accurately, calibration curve is set-up by plotting absorbance versus concentration of KI.

Procedure

[I] Calibration curve

Prepare the following mixtures in test tubes:

Test tube no.	Volume of (cm^3) 0.01M KI	Saturated $K_2S_2O_8$	Absorbance at λ_{max} = ______
1	2.5	17.5	
2	5.0	15.0	
3	7.5	12.5	
4	10.0	10.0	
5	12.5	7.5	

Shake each test tube to liberate iodine completely. Using the densest coloured solution, select λ_{max}. Note down the absorbance for each solution. Plot a graph of absorbance versus equilibrium concentration of KI solution. The concentration of KI solution is equivalent to the concentration of iodine liberated. This graph is called a calibration curve.

[II] Investigation of the reaction

Pipette out 10 cm^3 of 0.05M KI solution and 10 cm^3 of 0.05M $K_2S_2O_8$ solution separately in two test tubes. Keep them in a water bath for half an hour to attain room temperature.

Mix the two solutions and start the stopwatch. Shake well and take the solution in a colorimetric cell. Measure the absorbance of the solution after 2, 5, 10, 15 and 20 minutes at the λ_{max} selected without taking the cell out of the colorimeter.

Plot a graph of absorbance versus time. For a particular value of absorbance at some time t, the value of x can be determined from the calibration curve. Calculate the rate constant from the equation

$$K = \frac{1}{t} = \frac{x}{a(a-x)}$$

where a is the initial concentration of KI

x is the concentration of iodine at time t.

A plot of $x/(a-x)$ will have slope as aK. Thus the rate constant K is slope/a.

Observations and calculations

Determination of λ_{max}

The solution in test tube no.5 was chosen for determination of λ_{max}.

Wavelength λ (nm)	Absorbance
420	
470	
535	
620	
660	

Calibration curve

Test tube	Volume of (cm^3) 0.01M KI	Volume of (cm^3) Saturated $K_2S_2O_8$	Absorbance at λ_{max} = 420 nm
1	2.5	17.5	
2	5.0	15.0	
3	7.5	12.5	
4	10.0	10.0	
5	12.5	7.5	

Investigation of the reaction

Initial concentration of KI = a = ______ M

Time t (min)	Absorbance at λ_{max} = 420 (nm)	Conc. of I_2 × (mol)	$a - x$ (mol)	$x/(a - x)$	K = $1/t \cdot x/(a - x)$ $mol^{-1}\ dm^3\ min^{-1}$
2					
5					
10					
15					
20					

Mean K = _____ $mol^{-1}\ dm^3\ min^{-1}$ = _____ $mol^{-1}\ dm^3\ s^{-1}$

From the graph of $x/(a - x)$ versus time t,

Slope = ______

K = slope/a = _____ $mol^{-1}\ dm^3\ min^{-1}$ = _____ $mol^{-1}\ dm^3\ s^{-1}$

Result

The rate constant of the reaction between potassium iodide and potassium persulphate is

K = __________ $mol^{-1}\ dm^3\ s^{-1}$ (by calculations)

K = __________ $mol^{-1}\ dm^3\ s^{-1}$ (graphically).

Appendix

Review Questions

1. **What is the role of sodium acetate in the kinetic studies of the reaction between acetone and iodine?**
2. **What is the physical significance of the rate constant?**
3. **What is the role of potassium iodide in the kinetic studies of the autocatalytic reaction between potassium permanganate and oxalic acid?**
4. **What is the significance of T_0 and T_∞ in Experiment 5.1? If the reactants are mixed at a given time but the titration is done after 5 minutes, can that reading be considered a zero reading?**

Tables

Table 5.1: Kinetic data for first-order reactions

Reaction	Phase	Temperature (°C)	Rate constant k / s^{-1}	Half-life $t_{1/2}$
$2N_2O_5 \rightarrow 4NO_2$	g	25	3.38×10^{-5}	5.70 h
$+ O_2$	HNO_3(l)	25	1.47×10^{-6}	131 h
	Br_2(l)	25	4.27×10^{-5}	4.51 h

Reaction	Phase	Temperature (°C)	Rate constant k / s^{-1}	Half-life $t_{1/2}$
$C_2H_6 \rightarrow 2CH_3$	g	700	5.36×10^{-4}	21.6 min
Cyclopropane → propene	g	500	6.71×10^{-4}	17.2 min
$CH_3N_2CH_3 \rightarrow C_2H_6 + N_2$	g	327	3.4×10^{-4}	34 min
Sucrose → glucose + fructose	aq(H^+)	25	6.0×10^{-5}	3.2 h

Table 5.2: Kinetic data for second-order reactions

Reaction	Phase	Temperature (°C)	Rate constant k / (L mol-1 s-1)
$2NOBr \rightarrow 2NO + Br_2$	g	10	0.80
$2NO_2 \rightarrow 2NO + O_2$	g	300	0.54
$H_2 + I_2 \rightarrow 2HI$	g	400	2.42×10^{-2}
$D_2 + HCl \rightarrow DH + DCl$	g	600	0.141
$2I \rightarrow I_2$	g	23	7×10^{9}
	hexane	50	1.8×10^{10}
$H^+ + OH^- \rightarrow H_2O$	water	25	1.35×10^{11}
	Ice	-10	8.6×10^{12}

Expected Experimental Results

Experiment 5.1

The energy of activation of hydrolysis of methyl acetate catalysed by hydrochloric acid is ___ kJ mol^{-1}.

Experiment 5.2

The observed results show that the rate constant of the reaction increases with increase in ionic strength.

For the equation $\log K = \log K_a + 2\sqrt{I}$ to be valid, the slope should be 2. This expected value may not be obtained because according to the Debye-Huckel theory the equation holds good for only very dilute solutions ($\approx 10^{-4}$ M), which are not used in this experiment. The ice used to quench the reaction mixture should be made with distilled water otherwise the ionic strength of the solution increases as the ice melts due to dissolved salts which go into solution.

Experiment 5.3

Order of the reaction between acetone and iodine with respect to
acetone = 1; sulphuric acid = 1; iodine = 0.

Experiment 5.4

The second order rate constant for the hydrolysis of ethyl acetate by sodium hydroxide is of the order of 10^{-3} mol^{-1} dm^{3} min^{-1} in the temperature range 28 – 35 °C.

Experiment 5.5

The order of the reaction with respect to potassium permanganate is 2.

Experiment 5.6

The rate constant of the reaction between potassium iodide and potassium persulphate is of the order of 10^{-2} mol^{-1} dm^{3} min^{-1} in the temperature range 28 – 35°C.

Hints to Review Questions

1. Sodium acetate is added to arrest the reaction between acetone and iodine by raising the pH of the solution.
2. The rate constant is a measure of the number of molecules reacting in one unit of time.
3. Potassium permanganate oxidises potassium iodide to iodine. The released iodine is titrated with standard sodium thiosulphate. The volume of sodium thiosulphate required is thus an indirect estimation of the amount of potassium permanganate in the aliquot removed from the reaction mixture, and hence of the extent of completion of the reaction.
4. T_0 represents the concentration of HCl in the solution and serves as a baseline reading. T_∞ represents the total amount of acid in the solution after complete hydrolysis of methyl acetate. If the reactants are mixed at a given time but the first titration is done after 5 minutes, that reading can also be considered as a baseline reading.

Further Exploration through the Internet

http://www.cchem.berkeley.edu/~chem130a/sauer/outline/kinetics.html
http://users.erols.com/merosen/kinetics.htm
http://www.sparknotes.com/chemistry/kinetics/.dir/

Ernest Otto Beckmann

Ernest Otto Beckmann was born in 1853 in Solingen, Germany. He started his career as an assistant to a pharmacist, and eventually studied pharmacy and chemistry at Leipzig. He taught pharmacy and chemistry at the Technische Hochschule in Brunswick, before moving back to Leipzig as an assistant to Wislicenus.

Beckmann was interested in the physical and chemical properties of oximes, a class of organic compounds containing the –C=N–OH group. He discovered a means of obtaining oximes from ketones and then rearranging them to form amides. This reaction in organic chemistry is now known as Beckmann rearrangement, in honour of its discoverer. The rearrangement is very important in the industrial manufacture of Nylon-6. The current major process to produce caprolactam involves liquid phase Beckmann rearrangement of cyclohexanone oxime produced through reaction between hydroxylamine and cyclohexanone.

Around 1888, Beckmann's interest in oximes lead him to invent a number of ingenious ways of measuring their colligative properties. Based on the theoretical work of the French scientist, Francois-Marie Raoult, he developed a differential thermometer that could accurately measure the very small temperature changes in the calculation of molecular weight by freezing point depression or boiling point elevation. This thermometer also bears his name.

Beckmann was appointed to the post of Professor of Physical Chemistry at Leipzig. He later became the Director of the Laboratory of Applied Chemistry at Leipzig. He spent the final years of his professional career as the Director of the prestigious Kaiser Wilhelm Institute of Physical Chemistry and Electrochemistry in Berlin-Dahlem. He died in 1923.

6

Cryoscopy

6.1 Introduction

At a given temperature, a substance exists in the phase for which its chemical potential is the lowest at that temperature. For example, at 1 atm pressure, ice is the most stable form of water below 0°C but above 0°C liquid water is more stable. This is because the chemical potential of ice is less than that of liquid water below 0°C but the reverse is true above 0°C. The temperature at which the chemical potential of the two phases is equal is called the *phase transition temperature*.

Figure 6.1 shows how the chemical potential of a substance varies with temperature. The temperature at which the chemical potential of the solid and liquid phases are equal is the freezing point T_f of the substance while the temperature at which the chemical potentials of the liquid and gaseous phases are equal is the boiling point T_b of the substance.

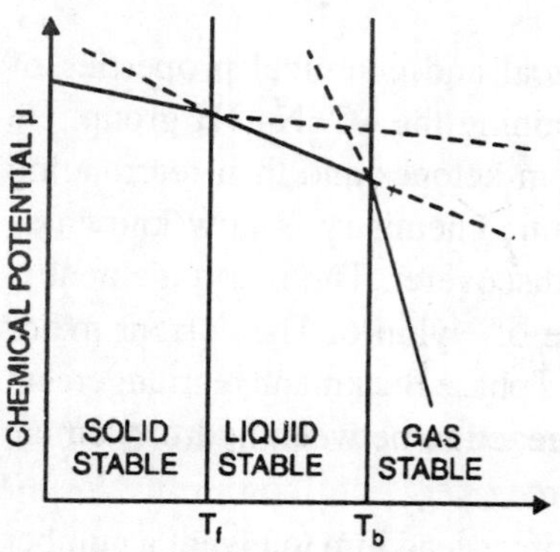

Figure 6.1: Variation in the chemical potential with temperature.

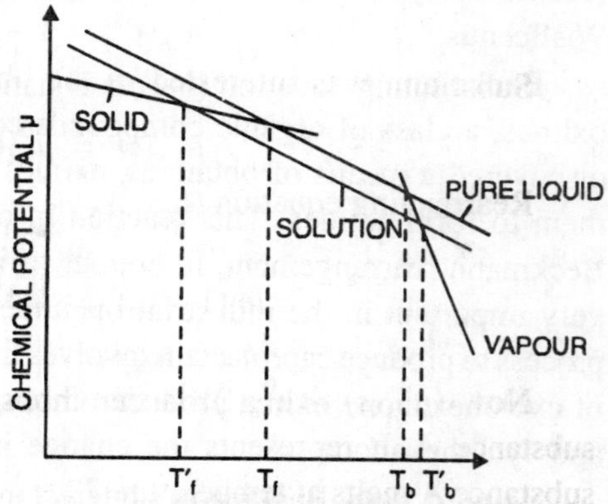

Figure 6.2: Variation in the chemical potential of the solution with temperature.

When a non-volatile solute is dissolved in a solvent, the chemical potential of the solvent is lowered. If μ^0_A is the chemical potential of the pure solvent A then the chemical potential μ_A due to the presence of a solute B is given by

$$\mu_A = \mu^0_A + RT \ln x_A \quad (6.1)$$

where x_A is the mole fraction of the solvent.

Since x_A is a fraction, $\ln x_A$ is a negative quantity. Thus,

$$\mu_A < \mu^0_A \quad (6.2)$$

The variation in the chemical potential of the solution with temperature is shown in Figure 6.2. The solute does not affect the solid or vapour phase of the solvent because it is not present in either. Figure 6.2 shows that as a result of the lowering in chemical potential of the solvent the boiling point is raised ($T_b' > T_b$) while the freezing point is lowered ($T_f' < T_f$). The extent of depression in the freezing point is more than the extent of elevation in the boiling point.

It can be shown that the elevation in boiling point and the depression in freezing point depend only on the number of solute particles present in solution and not on their nature. Such properties of solutions are called *colligative properties.* Other examples of colligative properties are osmotic pressure and relative lowering of vapour pressure of a pure solvent.

6.2 The Depression in Freezing Point

Consider a pure solvent A in which a non-volatile solute B is dissolved. According to Figure 6.2, at the new freezing point T_f' the chemical potential of the pure solvent in the frozen state, $\mu^0_A(s)$, will be equal to the chemical potential of the solution.

$$\mu^0_A(s) = \mu^0_A(\text{l}) + RT_f' \ln x_A \tag{6.3}$$

The mole fraction of the solvent x_A can be expressed in terms of the mole fraction of the solute x_B as follows

$$x_A = 1 - x_B \tag{6.4}$$

Substituting equation (6.4) in equation (6.3),

$$\mu^0_A(s) = \mu^0_A(\text{l}) + RT_f' \ln (1 - x_B) \tag{6.5}$$

Rearranging equation (6.5),

$$\ln (1 - x_B) = \frac{\mu^0_A(s) - \mu^0_A(l)}{RT_f'} \tag{6.6}$$

Now ($\mu^0_A(s) - \mu^0_A(\text{l})$) is the Gibbs free energy of fusion $\Delta_{\text{fus}}G$ of the substance A. It represents the change in free energy that occurs when the substance A melts at temperature T_f'.

$$\ln (1 - x_B) = \frac{\Delta_{fus} G}{RT_f'} \tag{6.7}$$

The change in Gibbs free energy can be expressed in terms of the change in enthalpy and the change in entropy at temperature T_f'.

$$\Delta_{\text{fus}}G = \Delta_{\text{fus}}H - T_f'\Delta_{\text{fus}}S \tag{6.8}$$

Substituting equation (6.8) in equation (6.7),

$$\ln (1 - x_B) = \frac{\Delta_{fus} H}{RT_f'} - \frac{\Delta_{fus} S}{R} \tag{6.9}$$

When $x_B = 0$ the freezing point is that of the pure liquid form T_f. Also, $ln(1 - x_B)$ becomes $ln\ 1$ which is zero. Substituting $x_B = 0$ in equation (6.9),

$$0 = \frac{\Delta_{fus}H}{RT_f} - \frac{\Delta_{fus}S}{R} \tag{6.10}$$

Subtracting equation (6.10) from (6.9),

$$ln\,(1 - x_B) = \frac{\Delta_{fus}H}{R}\left(\frac{1}{T_f'} - \frac{1}{T_f}\right) \tag{6.11}$$

If it is assumed that the solution is so dilute that $x_B << 1$ then

$$ln\,(1 - x_B) \approx -x_B$$

$$x_B = \frac{\Delta_{fus}H}{R}\left(\frac{1}{T_f} - \frac{1}{T_f'}\right) \tag{6.12}$$

$$x_B = \frac{\Delta_{fus}H}{R}\left(\frac{T_f' - T_f}{T_f T_f'}\right) \tag{6.13}$$

Since the difference between T_f and T_f' is very small, $T_f T_f' \approx T_f^2$. Equation (6.13) then becomes

$$x_B = \frac{\Delta_{fus}H}{R}\left(\frac{\Delta T_f}{T_f^2}\right) \tag{6.14}$$

where $\Delta T_f = T_f' - T_f$

Equation (6.14) shows clearly that the depression in the freezing point does not depend on the nature of the solute but only on its mole fraction. However, the extent to which the depression in the freezing point of a given solvent occurs depends on the properties of the solvent.

6.3 Determination of Molar Mass of the Solute

The depression in the freezing point of a liquid caused by the presence of a solute can be used to determine the molecular weight of the solute.

The mole fraction x_B of the solute can be expressed in terms of the number of moles of the solvent n_A and solute n_B present.

$$x_B = n_B/(n_A + n_B) \tag{6.15}$$

If the solution is very dilute, $n_B << n_A$ and $(n_A + n_B) \approx n_A$.

$$x_B = n_B/n_A \tag{6.16}$$

The number of moles of a substance is the ratio of the actual weight of the substance present and its molecular weight.

$$n_B = W_B/M_B \text{ and } n_A = W_A/M_A \tag{6.17}$$

The molality of a solute in a solution is the number of moles of the solute present in 1000g of the solvent.

Molality $$m = \frac{n_B}{W_A}.1000 \qquad (6.18)$$

Substituting equations (6.17) and (6.18) in equation (6.16),

$$x_B = \frac{m\,W_A}{1000} \cdot \frac{M_A}{W_A}$$

$$x_B = \frac{M_A m}{1000} \qquad (6.19)$$

Substituting equation (6.19) in equation (6.14),

$$\frac{\Delta_{fus} H}{R} \frac{(\Delta T_f)}{T_f^2} = \frac{M_A m}{1000}$$

$$\Delta T_f = \frac{(M_A R T_f^2)}{1000\,\Delta_{fus} H} m \qquad (6.20)$$

$$\Delta T_f = K_f m \qquad (6.21)$$

where $$K_f = \frac{(M_A R T_f^2)}{1000\,\Delta_{fus} H} \qquad (6.22)$$

The constant K_f is known as the molal freezing point depression or cryoscopic constant for 1000g of the solvent. It may be defined as the depression in freezing point caused when one mole of the solute is present in 1000g of the solvent.

The cryoscopic constant for water may be calculated substituting the values $M_A = 18$ g mol^{-1}, $R = 8.314$ J K^{-1} mol^{-1}, $T_f = 273$ K and $\Delta_{fus}H = 6.02$ kJ mol^{-1}.

$$K_f = \frac{(18\,\text{g mol}^{-1}) \times (8.314\,\text{JK}^{-1}\text{mol}^{-1}) \times (273\,\text{K})^2}{(1000\,\text{g kg}^{-1}) \times (6.02 \times 10^3\,\text{J mol}^{-1})}$$
$$= 1.86\ \text{K kg mol}^{-1}$$

Equation (6.21) can also be written as

$$\Delta T_f = K_f \cdot \frac{1000\,W_B}{M_B W_A} \qquad (6.23)$$

Rearranging,

$$M_B = \frac{1000\,K_f W_B}{\Delta T_f\,W_A} \qquad (6.24)$$

Thus the molecular weight of a solute can be determined from the depression in freezing point caused when W_B g of the solute is dissolved in W_A g of the solvent, provided the cryoscopic constant of the solvent is known. This method of molecular weight determination is known as *cryoscopy*.

6.4 Abnormal Behaviour of Solutions

The colligative properties of a solution depend only on the number of particles in solution. Solutes, which undergo association or dissociation in solution, give rise to abnormal behaviour in their solutions.

For example, when one mole of sodium chloride is dissolved in water, it causes twice the expected degree of depression in the freezing point because it dissociates into two ions in solution. On the other hand, when one mole of benzoic acid is dissolved in benzene, only half the expected degree of depression in the freezing point takes place because benzoic acid associates to form a dimer in solution.

The ratio of the experimentally determined value of a colligative property, Δ_{obs}, to the theoretical value, Δ_{calc}, calculated on the basis of the expected behaviour of the solute, is known as the van't Hoff factor, *i*.

$$i = \Delta_{obs}/\Delta_{calc} \qquad (6.25)$$

Since colligative properties are inversely proportional to the molar mass of the solute (see equation 6.23),

$$i = \frac{\text{Molar mass determined from formula}}{\text{Molar mass determined experimentally}} \qquad (6.26)$$

Consider *N* molecules of an electrolyte present in solution. Each molecule dissociates to give *M* ions. If α is the degree of dissociation,

Number of molecules which dissociate = $N\alpha$

Number of ions produced by dissociation = $NM\alpha$

Number of molecules left undissociated = $N - N\alpha$

Therefore, total number of species present in solution = $MN\alpha + N - N\alpha$

From equation (6.25),

$$i = \frac{(MN\alpha + N - N\alpha)}{N}$$

$$= M\alpha + 1 - \alpha$$

$$i - 1 = \alpha(M - 1)$$

Therefore, the degree of dissociation α is given by

$$\alpha = \frac{(i-1)}{(M-1)} \qquad (6.26)$$

This relation is used to determine the degree of dissociation of an electrolyte in solution if the van't Hoff factor is known.

6.5 Determination of Activity Coefficient of an Electrolyte

The activity coefficient of an electrolyte can be obtained by Bjerrum relation:

$$\log \gamma_\pm = [-\int_0^m (j/m)\, dm] - j \qquad (6.27)$$

where $\gamma_\pm$ is the mean activity coefficient of the ions

m is the molality of the solute

j is a function introduced by Lewis and Randall, which is a measure of the deviation from ideality of an electrolyte, and is equal to zero when m is zero. It is defined by the relation

$$j = 1 - \frac{\Delta T_f}{\nu m K_f} \qquad (6.28)$$

where ν is the number of ions produced by dissociation of one molecule.

6.6 Experimental Determination of the Depression in Freezing Point

The experimental determination of the depression in freezing point is carried out using a special thermometer called the *Beckmann thermometer*. The Beckmann thermometer has a large bulb at its bottom that contains mercury. At the top of the capillary is a mercury reservoir. The capillary corresponds to a scale of about 5 to 6 °C. Each degree is graduated in 100 equal parts.

The apparatus for freezing point determination consists of a freezing point tube, *A*, about 15-20 cm long and 2.5-3 cm wide with a side inlet tube, *T*, and a cork through which a Beckmann thermometer, *B*, and a stirrer, S_1, pass. The side tube is for introducing the solute into the liquid inside the freezing point tube. The tube A is surrounded by a larger tube, *C*, the space between the two forming an air jacket. This ensures slow and uniform cooling of the liquid in the inner tube. Both the tubes are surrounded by a larger vessel, *D*, which is filled with a suitable freezing mixture (usually ice and common salt). The freezing mixture is kept in circulation by means of another wire stirrer S_2. Any water formed due to the melting of the ice is periodically siphoned out.

The Beckmann thermometer is extremely sensitive and can measure temperature changes in the range of five to six degrees within an accuracy of 0.01°C. It is specially designed to measure relative changes in temperature. The actual temperature of a system cannot be measured.

6.6.1 Adjustment of the Beckmann thermometer

The thermometer is mounted in a clamp so that it hangs in a beaker of water. The temperature should be the temperature at which the experiment is to be conducted. The mercury bulb is allowed to attain the temperature of the bath. Check if the top of the mercury thread lies on the scale. If it does, the thermometer is set. If it does not, there are two possibilities:

1. The mercury thread rests below the scale

This indicates too little mercury in the lower bulb. To correct this, mercury has to be transferred from the reservoir into the bulb. Place the thermometer in a bath whose temperature is high enough to cause the mercury to rise to the top and form a droplet at the end of the capillary. Invert the thermometer and tap it gently so that the mercury at the end of the capillary joining with the mercury in the reservoir. Turn the thermometer gently into its original upright position and again place the bulb into the bath. The mercury in the bulb will now draw in more mercury from the reservoir. After a few minutes the bulb would have attained the temperature of the bath. Very gently tap the upper end of the thermometer against the palm of the hand so as to cause the excess mercury to break off from the end of the capillary. Check the adjustment by placing the thermometer in the bath whose temperature corresponds to the temperature at which the experiment is to be performed. And see if the mercury thread lies on the scale. If the mercury is too low on the scale, repeat the above procedure. If the mercury stands above the scale, perform the procedure outlined below.

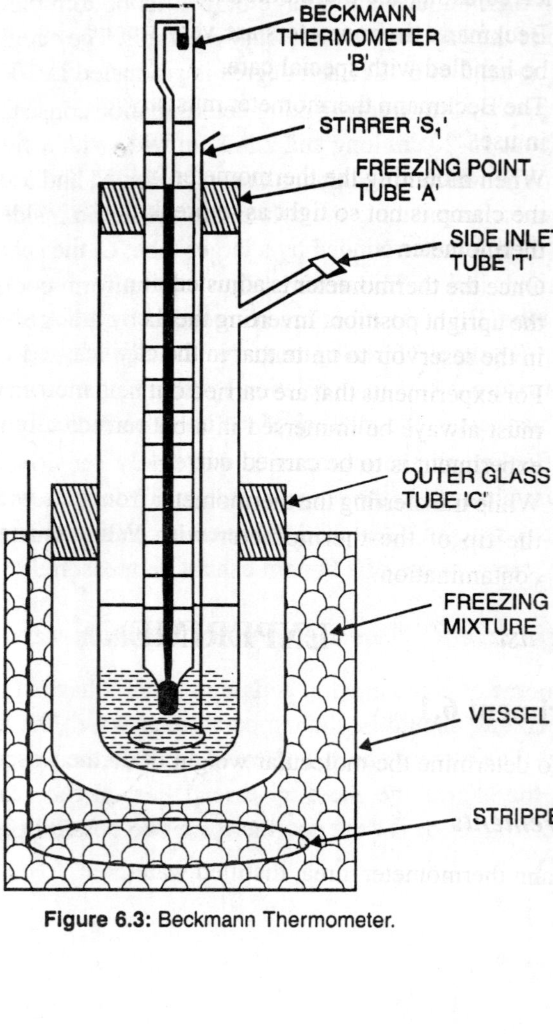

Figure 6.3: Beckmann Thermometer.

2. The mercury thread rests above the scale

This indicates that there is too much mercury in the lower bulb. To correct this, mercury has to be driven up from the bulb into the reservoir. Place the bulb of the thermometer in a bath whose temperature is just 2-3°C above the freezing point of the pure solvent being used. After waiting for several minutes gently tap the upper end of the thermometer to detach any droplet formed at the end of the capillary. Check the adjustment of the thermometer by immersing its bulb in the water bath whose temperature matches the temperature at which the experiment is to be performed.

Repeat procedures (a) and (b) until the mercury level rests on the upper part of the scale at the freezing point of the pure solvent being used.

6.6.2 Precautions

As mentioned earlier, the setting of the Beckmann thermometer requires a lot of patience and practice. The following precautions should always be kept in mind when adjusting the thermometer:

1. Beckmann thermometers are very fragile and expensive, and should be handled with special care.
2. The Beckmann thermometer must always be kept in its case when not in use.
3. When mounting the thermometer in the clamp, make sure the grip of the clamp is not so tight as to break the thin outer glass casing of the thermometer.
4. Once the thermometer is adjusted, it must be always be maintained in the upright position. Inverting the thermometer will cause the mercury in the reservoir to unite that in the capillary, thus spoiling the setting.
5. For experiments that are carried out below room temperature the bulb must always be immersed in a bath at the temperature at which the experiment is to be carried out.
6. While transferring the thermometer from one solution to another, wipe the tip of the thermometer with Whatman filter paper to avoid contamination.

EXPERIMENT

Experiment 6.1

Aim: To determine the molecular weight of urea cryoscopically.

Requirements

Beckmann thermometer, urea, distilled water, ice, freezing tube etc.

Procedure

Take 20g of water in a freezing tube. Adjust the Beckmann thermometer so that when immersed in a bath containing ice and water, the mercury is between 0 and –1. Now determine the freezing point of water by placing the tube in an ice-salt mixture and placing the Beckmann thermometer in it. When its temperature is about 1°C, warm the freezing tube under the tap. Add 0.2g of urea so that it forms a 1% solution and dissolve the urea. Record the freezing point once again. Continue by adding 2%, 3%, 4% and 5% solutions and determine the freezing point of each.

Plot a graph of w_1/w_2 versus ΔT_f where

w_1 is the amount of urea added in g

w_2 is the weight (in g) of water taken in the freezing tube

ΔT_f is the depression in freezing point

Calculate the slope of the graph and using the equation

Molecular weight of urea = slope × K_f × 1000

calculate the molecular weight of urea.

Observations and calculations

Freezing point of water on Beckmann thermometer T_f = ______ °C

Cryoscopic constant of water K_f = 1.86 kg mol^{-1} °C

Weight of pure water taken in freezing tube w_2 = ___ g

Wt. of urea added w_1 (g)	w_1/w_2	T (°C)	$\Delta T_f = T_f - T$ (°C)

From the graph of w_1/w_2 versus ΔT_f, slope = ______

Molecular weight of urea = slope × K_f × 1000 = ______

Result

Molecular weight of urea = ______ g mol^{-1}.

Experiment 6.2

Aim: To determine the activity coefficient of an electrolyte (e.g., HCl) in dilute solution by equilibrium freezing point depression method.

Requirements

Dewar flask, Beckmann thermometer, stirrer, burette, pipette, 1N HCl solution having other different molalities, ice, 0.2N NaOH.

Procedure

Put some finely crushed ice into a Dewar flask so that it is half-filled. Add enough ice-cold water to fill the space in between the ice pieces. The flask now contains a mixture of ice and water in equilibrium with each other and temperature of the mixture gives the true freezing point of water. The Beckmann thermometer is set so that it stands between 0 and – 2 on scale at ice point. Insert thermometer in flask and stir the ice and water until a steady reading is recorded. This is the freezing point of water. Close the third hole in the cork with cotton. Add 25 cm^3 ice-cold 1N HCl into the mixture to obtain a new equilibrium freezing point. Record the steady temperature. Remove about 25 cm^3 of this solution from the flask and draw into previously weighed conical flask. Let the solution be warmed to room temperature.

Remove the condensed moisture on the outside. Weigh and determine the weight of the solution. Titrate this against approximately 0.2N NaOH using phenolphthalein. Add 25 cm^3 of 1N HCl and determine the freezing point. Thus determine equilibrium freezing point for 6 different concentrations of HCl. Use approximately 10 cm^3 for last two solutions for analysis. Standardise NaOH with 0.1N succinic acid. Calculate the amount of solute present in the solution. Evaluate *j* and *j*/*m*. Plot *j*/*m* versus *m* and from this, calculate area below the curve. Evaluate the integral $\int_0^m j/m \, dm$.

Observations and calculations

Freezing point of water on Beckmann thermometer T_f = ______°C

Cryoscopic constant of water K_f = ______ kg mol^{-1} °C

Vol. of 0.1N HCl added (cm^3)	Freezing point T (°C)	Depression in freezing point $\Delta T_f = T_f - T$ (°C)	Vol. of HCl taken for titration V_1 (cm^3)	Wt. of empty flask w_1 (g)	Wt. of soln + flask w_2 (g)

Wt. of soln $w = w_2 - w_1$ (g)	Titre vol of 0.2N NaOH V_2 (cm^3)	$C_{HCl} = V_2 C_{NaOH} / V_1$ (N)	Moles of HCl $n = V_1 C_{HCl}/1000$	Wt. of HCl = 36.5n (g)	Wt. of water $w_{H_2O} = w - w_{HCl}$ (g)

Molality $m = (w_{HCl}/w_{H_2O}) \times (1000/36.5)$	$j = 1 - (\Delta T_f / \acute{\imath}_m K_f)$	j/m	$\int_0^m j/m \, dm$ (area under the curve in cm^2)	$\ln \gamma_\pm = [-\int_0^m (j/m)\, dm] - j$	$\gamma_\pm$

Result

The activity coefficient of the electrolyte HCl _______(increases/decreases/reamins the same) as its concentration increases.

Experiment 6.3

Aim: To determine the mean ionic activity coefficient of KCl and $BaCl_2$ by equilibrium freezing point depression method.

Requirements

Dewar flask, Beckmann thermometer, stirrer, burette, pipette, KCl solution of different molalities, ice.

Procedure

Put some finely crushed ice in a Dewar flask. Add enough ice cold water to cover the ice. The flask now contains a mixture of ice and water at equilibrium with each other. The temperature of this mixture gives the true freezing point of water. The Beckmann thermometer is set so that it reads between 0 and 1 on the scale at the ice point. Take 50 cm^3 of 1 M KCl solution in the Dewar flask and determine its freezing point. In a similar manner determine the freezing point of KCl solutions of 0.8, 0.6, 0.4, 0.2, 0.1 and 0.05 molality. Calculate *j* and *j*/*m* for each molality. Plot a graph of *j*/*m* versus m and calculate the activity coefficients by determining the area below the curve. Evaluate the integral $_0\square^m$ *j*/*m dm*.

Observations and calculations

Freezing point of water on Beckmann thermometer T_f = _____ °C

Cryoscopic constant of water K_f = _____ kg mol^{-1} °C

Molality of KCl *m* / (m)	Freezing point *T* (°C)	Depression in freezing point $\Delta T_f = T_f - T$ (°C)	$j = 1 - (\Delta T_f / V_m K_f)$

j/*m*	$_0\square^m$ *j*/*m dm* (area under the curve in cm^2)	$\ln \gamma_\pm = [-\,_0\square^m (j/m)\ dm] - j$	$\gamma_\pm$

Result

The activity coefficient of KCl solution ______ (increases/decreases/remains the same) as its molality increases.

Appendix

Review Questions

1. For what purpose are zero-bottom and zero-top Beckmann thermometers used?
2. What is the least count of the Beckmann thermometer?
3. Why is there a small mercury reservoir at the top of the Beckmann thermometer?
4. What are the other solute-solvent pairs that can be used in cryoscopy?

Table 6.1: Cryoscopic constants

Compound	*Freezing Point (°C)*	K_f / *(K kg mol^{-1})*
Water	0.0	1.853
Formic acid	8.27	2.77
Formamide	2.55	3.85
Acetic acid	16.66	3.90
Acetamide	80.00	4.04
Dimethylsuplhoxide	18.54	4.07
p- Xylene	13.263	4.3
1,4-Dioxane	11.8	4.63
Benzene	5.533	5.12
o-cresol	30.944	5.60
Dibenzyl ether	3.60	6.27
N-Methylacetamide	30.55	6.65
Nitrobenzene	5.76	6.852
Naphthalene	80.290	6.90
p-Cresol	34.739	6.96
Phenol	40.90	7.40
Diphenyl ether	26.87	7.88
2-Methyl-2-propanol	25.82	8.37
2,2-Dimethyl-1-propanol	52	11.0
1,2-Dibromoethane	9.79	12.5
Bromoform	8.05	14.4
Bicyclohexane	3.63	14.52
Antimony(III) chloride	73.4	17.95
Succinonitrile	57.88	18.26
Cyclohexane	6.54	20.0
1,1,2,2-tetrabromoethane	0.0	21.7
Carbon tetrachloride	– 22.95	29.8
Camphor	178.75	37.7
Cyclohexanol	6.544	39.3
Sulpholane	–30.9	64.1

Table 6.2: Ebullioscopic constants

Compound	*Boiling Point (°C)*	K_b / (K kg mol^{-1})
Water	100.00	0.515
Methanol	64.70	0.785
Ethanol	78.29	1.160
1-propanol	97.20	1.59
Acetone	56.29	1.71
Diethyl ether	34.55	1.824
Methyl acetate	56.323	2.061
Dimethoxymethane	42.30	2.125
2-Butanone	79.64	2.28
Carbon disulphide	46.225	2.35
Benzene	80.100	2.53
Acetic acid	117.9	2.530
Ethyl acetate	77.114	2.583
Pyridine	115.256	2.710
Cyclohexane	80.725	2.75
Piperidine	106.40	2.84
Aniline	184.40	3.22
N,N-dimethylacetamide	166.1	3.22
1,4-Dioxane	101.320	3.270
Toluene	110.625	3.29
Heptane	98.427	3.43
Propionic acid	140.83	3.51
Phenol	181.839	3.60
Chloroform	61.152	3.62
Chlorobenzene	131.687	4.15
Carbon tetrachloride	76.75	4.48
Camphor	207.42	5.611
Naphthalene	217.955	5.80
Bromobenzene	155.908	6.26
1,2-Dibromoethane	131.36	6.608

Expected Experimental Results

Experiment 6.1

The molecular weight of urea is 60 g mol^{-1}.

Experiment 6.2

The activity coefficient of the electrolyte HCl decreases as its concentration increases.

Experiment 6.3

The activity coefficient of KCl solution decreases as its molality increases.

For all three experiments, most errors arise from incorrect adjustment of the Beckmann thermometer. See Section 6.6. The solutions taken for the experiment should always be ice-cold.

Hints to Review Questions

1. Zero-top Beckmann thermometers are used to measure the depression in the freezing whereas zero-bottom Beckmann thermometers are used to measure the elevation in boiling point.
2. See Section 6.6.
3. See Section 6.6.1
4. See Table 6.1.

Further Exploration through the Internet

http://www.unmc.edu/PharmSciences/wwwcourse/colligative/sld002.htm
http://genomics.lbl.gov/~chem130A/ClassNotes/03_21_01/03_21_01.pdf

John Tyndall

John Tyndall was born in Ireland in 1820. He started his career in 1839 as an ordnance surveyor in the Irish army and then became a railway engineer in 1844. In 1847 he turned his interest to chemical research and initially studied the magnetic properties of crystals. In 1853 he was made a Fellow of the Royal Society and was appointed to the chair of natural philosophy at the Royal Institution, London. Michael Faraday was a colleague and in 1857 Tyndall succeeded him as Superintendent of the institution.

In 1859 Tyndall began studying the ability of gases to absorb and radiate heat. He constructed a spectrophotometer to measure the absorption of atmospheric gases. He concluded that water vapour is the strongest absorber among them, and speculated on its role in climate control.

Tyndall studied the scattering of light by dust and large molecules. He demonstrated that the blue colour of the sky is due to the scattering of sunlight by atmospheric dust. This scattering effect was named after him.

Tyndall had the ability to communicate difficult scientific concepts to the layman in a simple manner. He was a prolific writer and undertook many lecture tours. He penned a number of papers and essays on his scientific discoveries, including *Heat as a Mode of Motion* (1862), *On Radiation* (1865), *Forms of Water* (1872), and *Six Lectures on Light* (1873). He also wrote on more general topics like *Faraday as a Discoverer*, *Natural Philosophy in Easy Lessons*, *On the Scientific Use of the Imagination*, and *Fragments of Science*. Besides this, he also published works on religion, mountaineering and travel. He died in 1893 in England.

7
The Colloidal State

7.1 Introduction

When the size of the solute particles dispersed in a solvent is between 1 nm and 100 nm, the solute is said to be in a *colloidal state* and the solution is said to be a *colloidal solution*. Solutions in which the solute particles have size less than 1 nm are called *true solutions*. True solutions are homogeneous in nature. Hence they neither scatter light nor can they be filtered. Colloidal solutions appear homogeneous to the naked eye but they scatter light, thus indicating their heterogeneity. They can be filtered using filter papers with fine pores and made to settle by centrifugation.

The size of the colloidal particle may result from the aggregation of several smaller atoms and molecules. For example, a colloidal particle of gold is made up of several gold atoms clustered together while that of sulphur is made up of a thousand odd S_8 molecules. It may also happen that the colloidal particle is itself a big molecule (generally called *macromolecule*). The molecular mass of a macromolecule may range from a thousand odd to a few millions. Macromolecules are so large in size that their true solutions are actually colloidal in state.

7.2 Colloidal Solutions

Based on the nature of the solute and the medium in which it is dispersed, eight types of colloidal solutions are possible. They are shown in Table 7.1. Note that a colloidal solution of gas in gas is not possible because two gases are always completely miscible and always form a true solution.

Table 7.1: Various types of colloidal dispersions

Solute	*Dispersion medium*	*Name*	*Examples*
Solid	Solid	Solid sol	Porcelain, coloured glass, alloys
Solid	Liquid	Sol	Indian ink, paints, white of an egg
Solid	Gas	Aerosol	Smoke, volcanic dust
Liquid	Solid	Gel	Cheese, jellies
Liquid	Liquid	Emulsion	Milk, butter
Liquid	Gas	Aerosol	Cloud, mist, fog
Gas	Solid	Solid foam	Cake, dough, pumice stone, biscuit
Gas	Liquid	Foam	Whipped cream, foam

Sols can be classified as *lyophilic* (solvent loving) or *lyophobic* (solvent hating) according to the interaction between the solid and liquid phases.

Lyophilic sols are generally colloidal solutions of polymeric molecules like gelatine or starch in water, rubber in benzene, and cellulose acetate in acetone. Lyophilic sols are stable due to the strong, favourable solute-solvent interactions.

Lyophobic colloids are generally substances that are highly insoluble in the dispersing medium. Special methods are employed to prepare lyophobic sols. These methods can be broadly divided into two groups:

A. Condensation Methods

The particles present in a true solution are made to grow to colloidal dimensions by chemical or physical means. A few examples of this technique are:

1. **Bredig's method:** This is a classic method of producing a metal sol by passing an electric arc between the electrodes of the desired metal immersed in water. The vapourised metal forms aggregates of colloidal size.
2. **Hydrolysis:** A deep red sol of ferric hydroxide is obtained by adding ferric chloride solution to a beaker of boiling water.

$$FeCl_3 + 3H_2O \rightarrow Fe(OH)_3 + 3HCl$$

3. **Double decomposition:** A yellow sol of arsenious sulphide can be prepared by passing hydrogen sulphide gas though a dilute solution of arsenious oxide in water.

$$As_2O_3 + 3H_2S \rightarrow As_2S_3 + 3H_2O$$

B. Dispersion Methods

Bigger particles of a substance are broken down into smaller particles of colloidal dimensions in the presence of the dispersion medium. Two main examples of this process are:

1. **Grinding**: The solid is ground in a ball mill with the dispersing medium. Aquadag (a graphite lubricant) and Indian ink (lampblack suspended in water) are prepared in this way.
2. **Peptisation**: When an electrolyte is added to a precipitated salt, the fine particles of the precipitate adsorb ions from the added electrolyte and form a colloidal solution. The electrolyte added is called a peptising agent or dispersing agent.

   For example, when a small quantity of ferric chloride solution is added to freshly precipitated and well-washed ferric hydroxide, a reddish brown, positively charged sol is obtained.

$$\underset{\text{precipitate}}{Fe(OH)_3} + \underset{\text{ions adsorbed}}{Fe^{3+}} \rightarrow \underset{\text{sol particles}}{Fe(OH)_3.Fe^{3+}}$$

   A precipitate of Prussian blue can be prepared by mixing dilute solutions of ferric chloride and potassium ferrocyanide.

$$K_4[Fe(CN)_6] + FeCl_3 \rightarrow KFe[Fe(CN)_6]\downarrow + 3KCl$$

If oxalic acid solution is added as a peptising agent, this precipitate passes into colloidal solution.

7.2.1 Purification of colloidal solutions

Colloidal solutions prepared by chemical methods generally contain excessive quantities of electrolytes and other soluble impurities. Though small quantities of these electrolytes are required for stabilising lyophobic sols, their concentration should be reduced to a minimum because they cause precipitation of the sol on standing. The methods employed for this are:

1. Dialysis

Porous membranes like parchment and cellophane retain particles of colloidal size while allowing ionic impurities to pass through them. The impure sol is enclosed in a parchment or cellophane bag and suspended in running water in a trough. The soluble impurities diffuse out gradually, leaving the pure sol behind. This process is called *dialysis*. It can be speeded up by using hot water instead of cold (*hot dialysis*) and by dipping positive and negative electrodes in the water surrounding the bag (*electro dialysis*), as shown in Figure 7.1.

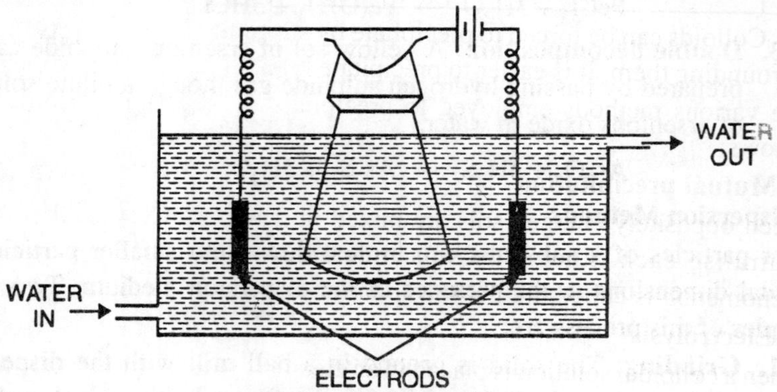

Figure 7.1: Electro dialysis.

2. Ultra filtration

The pore sizes of ordinary filter papers are too large to retain colloidal particles. The pore size can be reduced by impregnating with collodion solution. (Preparation of collodion: Cellulose is nitrated using a mixture of concentrated nitric and sulphuric acids to obtain pyroxylin, which is dissolved in ether or alcohol.) These ultrafilters retain the sol particles in the form of slime, which can be resuspended in water to get the sol.

3. Ultracentrifugation

The sol is placed in high-speed centrifugal machines that have 15,000 or more revolutions per minute. The colloidal particles settle down. This slime can be resuspended in water to get the sol.

7.2.2 Precipitation of colloids

Colloidal solutions have extraordinary stability. A gold sol prepared in 1857 by Michael Faraday is on display in the British Museum, London and shows no signs of ever settling down. The secret of the stability of colloidal solutions lies in the charge carried by each colloidal particle. Since all colloidal particles of a given sol carry the same charge, they repel each other. This prevents them from coagulating and settling down. Table 7.2 lists some common sols with the type of charge on their particles.

Table 7.2: Some common sols

No.	*Positively charged*	*Negatively charged*
1	Oxides, e.g., TiO_2 sol	Metals, e.g., Cu, Ag and Au sols
2	Metallic hydroxides, e.g., $Fe(OH)_3$, $Cr(OH)_3$ and $Al(OH)_3$ sols	Metallic sulphides, e.g., As_2S_3, CdS and Sb_2S_3 sols
3	Basic dyestuffs, e.g., methylene blue sol	Acid dyestuffs, e.g., Eosin, Congo red
4	Haemoglobin	Sols obtained from gums, starch, kaolin, charcoal etc.

Colloids can be forced to precipitate by stripping off the stabilising charge surrounding them. It is easier to precipitate lyophobic sols than lyophilic sols. The various methods employed to precipitate hydrophobic colloids are as follows:

1. Mutual precipitation

When oppositely charged sols are mixed in equal proportions, their charges neutralise each other. The sols thus coagulate and settle down. This phenomenon is called mutual precipitation.

2. Electrolysis

When a colloidal solution is placed in an electric field, its particles are attracted towards the electrode that carries a charge opposite to their own. Given sufficient time, the colloidal particles actually touch the electrode, lose their charge and coagulate.

3. Addition of electrolytes

When an electrolyte is added to a sol, the charge on the sol is neutralised by the opposite charge on the ions of the electrolyte. Hydrophobic sols like arsenious sulphide and ferric hydroxide are quite easily precipitated by the addition of small amounts of an electrolyte.

The *Hardy-Schulze Law* states that the quantity of the electrolyte required to coagulate a fixed amount of a colloid depends on the valency of the ion bearing a charge opposite to that of the colloidal particles.

For example, to precipitate the negatively charged arsenious sulphide sol, divalent cations are about 70 times as effective as monovalent ions, while trivalent ions are 500 times as effective. The effectiveness of an electrolyte to bring about precipitation of a sol is expressed in terms of its flocculation value, which is defined as the minimum concentration of the electrolyte which is able to cause coagulation or flocculation of a sol. Flocculation values are usually expressed in units of millimoles of the electrolyte per litre of the sol. The flocculation values of some common electrolytes for As_2S_3 and $Fe(OH)_3$ sols are given in Table 7.3.

Table 7.3: Flocculation values (in millimoles of the electrolyte per litre of sol) of some electrolytes for As_2S_3 and $Fe(OH)_3$ sols

Arsenious sulphide			*Ferric hydroxide sol*		
Electrolyte	*Cation*	*Flocculation value*	*Electrolyte*	*Anion*	*Flocculation value*
NaCl	Na^+	52	KCl	Cl^-	102
KCl	K^+	52	KBr	Br^-	134
KNO_3	K^+	50	KNO_3	NO_3^-	132
$MgSO_4$	Mg^{2+}	0.82	$KBrO_3$	BrO_3	133
$BaCl_2$	Ba^{2+}	0.69	K_2CrO_4	CrO_4^{2-}	0.325
$ZnCl_2$	Zn^{2+}	0.68	K_2SO_4	SO_4^{2-}	0.210
$AlCl_3$	Al^{3+}	0.093	$K_3[Fe(CN)_6]$	$[Fe(CN)_6]^{3-}$	0.096

4. Persistent dialysis

Traces of electrolytes are required for the stability of a colloidal solution. On prolonged dialysis, even these traces are removed and the colloid precipitates.

7.3 Micelles

Lyophilic sols are generally formed by polymeric molecules. However, there exist certain polymer molecules, which have both lyophilic and lyophobic groups present. Due to their dual character, they are called amphithetic.

For example, sodium stearate dissociates in solution to give Na^+ ions and $C_{17}H_{35}COO^-$ (stearate) ions. The stearate ion has a long hydrocarbon chain that is hydrophobic, but it also has a carboxylate group (COO^-) at the head of the chain that is hydrophilic. When present in dilute solutions the hydrocarbon chains tend to migrate to the surface, where their unfavourable interaction with the water molecules is minimum. At higher concentrations, they aggregate spontaneously to form a micelle, which is of colloidal dimensions. A micelle is a roughly spherical structure in which about a hundred stearate ions group together in such a way that their hydrocarbon tails are in the interior of the sphere while the ionic heads face outwards. This is shown in Figure 7.2. The ionic groups have strong, favourable interaction with water molecules. This stabilises the micelle. The highly charged micelle binds a fairly large number of positive ions to its surface as counter ions, which reduces its charge considerably. At even higher concentrations, the shape of the micelles change from flattened spheres to rod-like or lamellar.

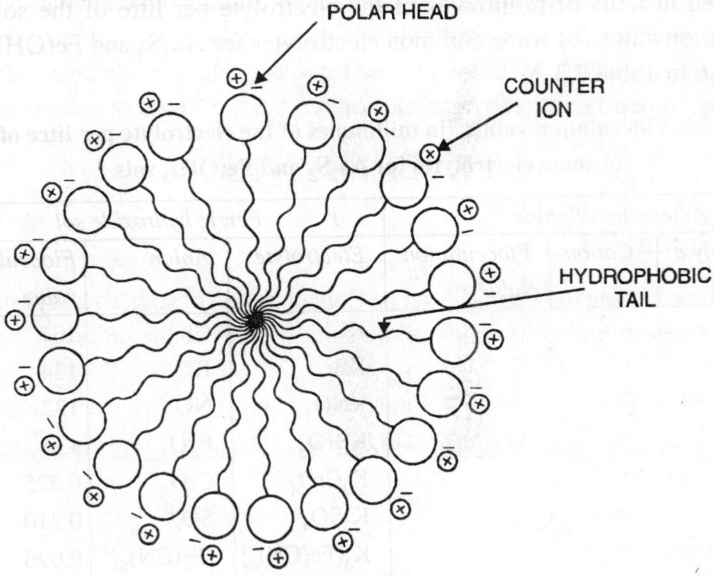

Figure 7.2: Micelle.

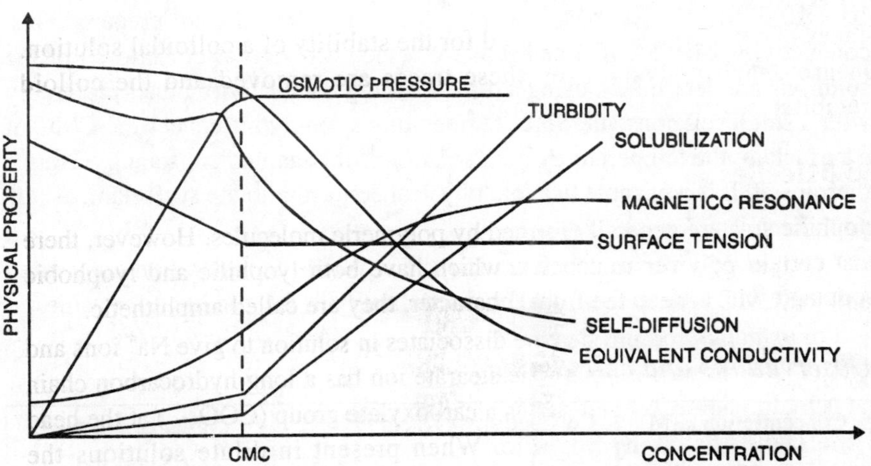

Figure 7.3: Discontinuity in the physical properties of a solution at the CMC.

Micelles are formed only above a certain minimum concentration and above a minimum temperature, both of which are characteristics of the system. The minimum concentration is called *Critical Micelle Concentration* (CMC) and the minimum temperature is called *Kraft temperature*. At the CMC there is a sharp discontinuity in physical properties of the solution like molar conductivity, osmotic pressure and surface tension, as shown in Figure 7.3.

7.3.1 Industrial importance of micelles

An important characteristic of micelles is that their interior is hydrocarbon in nature. It can dissolve similar hydrocarbon matter like fat, grease and oil. At the same time, the outer surface of the micelle is ionic, which makes it water-soluble. Micelles are thus used extensively in biology and industry for their solubilising capacity. Micelles are used as drug carriers, cleansers, in froth flotation, in petroleum recovery, and for synthesis of organic compounds.

EXPERIMENTS

Experiment 7.1

Aim: To determine the Critical Micelle Concentration of sodium lauryl sulphate from the measurement of conductivities at different concentrations.

Requirements

Digital conductivity meter, jacket cell, thermostat at 25°C, sodium lauryl sulphate *AR*.

Procedure

Solutions of sodium lauryl sulphate in conductivity water having concentrations 10×10^{-3} M, 9×10^{-3} M, 8×10^{-3} M, ...1×10^{-3} M are prepared. The conductivity k (in $S\ cm^{-1}$) and molar conductivity $\wedge$ (in $S\ cm^2\ mol^{-1}$) of the solutions are determined using a direct reading conductivity meter preferably with a small cell constant. Since temperature strongly influences the CMC of a surfactant, the temperature of the solution is maintained constant by using a thermostat. It is advisable to avoid physical contact with the surfactant, as it is an irritant.

Graphs of k versus concentration and $\wedge$ versus concentration are plotted and the CMC is detected from the discontinuity in the conductivity values.

Observations and calculations

Concentration c (M)	Conductivity k (S cm^{-1})	Molar conductivity $\wedge$ (S cm^2 mol^{-1})

From the graph of conductivity k versus concentration, CMC = mM

From the graph of molar conductivity $\wedge$ versus concentration, CMC = ____mM.

Result

The Critical Micelle Concentration of sodium lauryl sulphate is ______mol dm^{-3}.

Experiment 7.2

Aim: To determine the Critical Micelle Concentration of sodium lauryl sulphate from the measurement of the surface tension of the sodium lauryl sulphate/air interface by torsion balance method.

Requirements

Torsion balance, platinum ring, sodium lauryl sulphate AR.

Procedure

Solutions of sodium lauryl sulphate in conductivity water having concentrations 10×10^{-3} M, 9×10^{-3} M, 8×10^{-3} M, ...1×10^{-3} M are prepared.

The measurement of surface tension by the torsion balance method is rapid, simple and very accurate. The method depends essentially on the determination of the force required to detach a circular ring of platinum wire from the surface of a liquid when the angle of contact is 0°. Thoroughly clean the platinum wire with hot acid cleaning mixture, rinse with distilled water, and dry by momentarily heating in a luminous flame. Without touching the ring, suspend it from the hook on the beam. Turn the knob till the pointer is at zero and adjust the torsion of the wire so that the beam lies in the horizontal position and just clear of the arm. Raise the platform holding the watch glass and liquid until the liquid just touches the ring. Simultaneously turn the knob that twists the torsion wire and lower the platform by means of the screw underneath until the ring is torn from the surface. By this method the beam remains horizontal all the time, thus preventing high readings. Record the position of the pointer on the scale at which the ring is pulled free. Repeat several times for each liquid solution until concordant results are obtained.

Observations and calculations

Surface tension of water at room temperature = ______ dynes cm^{-1}

Concentration C (M)	Surface tension (dynes cm^{-1})			
	(I)	(II)	(III)	Mean

From the graph of surface tension versus concentration, CMC = ______mM.

Result

The Critical Micelle Concentration of sodium lauryl sulphate is ______mol dm^{-3}.

Experiment 7.3

Aim: To determine the flocculation value of As_2S_3 and $Fe(OH)_3$ sols using electrolytes having cations and anions of different valencies in order to test the validity of the Hardy-Schulze Rule.

Requirements

Colloidal solutions of As_2S_3 and $Fe(OH)_3$, 3 M KCl, 0.2M KCl, 5×10^{-3}M Na_2SO_4, 5×10^{-4}M $K_3[Fe(CN)_6]$, 2.5×10^{-3}M $CaCl_2$, 5×10^{-4}M $AlCl_3$, test tubes.

Procedure

A. Preparation of colloidal solutions

1. Arsenic sulphide

Add 0.1g of pure As_2O_3 in 150 cm^3 of boiling water. Reflux the solution for about 20 minutes. After the entire compound has dissolved, cool the mixture to room temperature and remove any undissolved arsenious oxide by simple filtration. Dilute to 500 cm^3 in a standard flask. Pass hydrogen sulphide gas through the solution for about 5 minutes. A yellow sol of arsenious sulphide will gradually form. Purge out excess hydrogen sulphide gas by bubbling nitrogen gas through the solution for about ten minutes. Filter the solution to remove any precipitate of arsenious sulphide that may have formed.

2. Ferric hydroxide

Pour 25 cm^3 of 2% ferric chloride solution gradually into 250 cm^3 of boiling distilled water. Hydrolysis of the chloride occurs instantly, forming a deep-red coloured sol of ferric hydroxide. Cool the sol to room temperature and purify by dialysis.

B. Determination of flocculation value

Take ten test tubes and number them from 1 to 10. Add by means of a burette 1, 2, 3, ... 10 cm^3 of 0.2M KCl to the test tubes 1 ... 10 respectively. Add distilled water to make the total volume in the test tubes 10 cm^3. To each test tube add 10 cm^3 of arsenious sulphide sol and stopper it. Gently mix the contents of the test tube by inverting it two or three times. Allow the test tubes to stand for two hours and then determine in which test tube there is just complete precipitation. Calculate the concentration of KCl in that test tube in mol dm^{-3}. This is the flocculation value of KCl for As_2S_3.

Repeat the above procedure for the determination of flocculation values of 2.5×10^{-3}M $CaCl_2$ and 5×10^{-4}M $AlCl_3$ for the arsenious sulphide sol.

Determine the flocculation values of 3 M KCl, 5×10^{-3} M K_2SO_4 and 5×10^{-4} M $K_3[Fe(CN)_6]$ for the ferric hydroxide sol in a similar way.

Observations and calculations

Flocculation value of KCl for As_2S_3 sol

Volume of 0.2 M KCl required to cause flocculation V_1 = ____ cm^3

Number of millimoles of KCl = 0.2 V_1 = ____

Volume of sol taken = 10 cm^3

Flocculation value = Number of millimoles of KCl per dm^3 of sol

$= 0.2 \times 100\ V_1 =$ ____

Flocculation value of $BaCl_2$ for As_2S_3 sol

Volume of 5×10^{-3} M BaCl required to cause flocculation $V_2 =$ ____ cm^3

Number of millimoles of BaCl = $5 \times 10^{-3}\ V_2 =$ ____

Volume of sol taken = 10 cm^3

Flocculation value = Number of millimoles of $BaCl_2$ per dm^3 of sol

$= 5 \times 10^{-3} \times 100\ V_2 =$ ____

Flocculation value of $AlCl_3$ for As_2S_3 sol

Volume of 5×10^{-4} M $AlCl_3$ required to cause flocculation $V_3 =$ ____ cm^3

Number of millimoles of $AlCl_3 = 5 \times 10^{-4}\ V_3 =$ ____

Volume of sol taken = 10 cm^3

Flocculation value = Number of millimoles of $AlCl_3$ per dm^3 of sol

$= 5 \times 10^{-4} \times 100\ V_3 =$ ____

Ratio of flocculation values = $V_1 : V_2 : V_3 = 1 :$ ____ $:$ ____

Flocculation value of KCl for $Fe(OH)_3$ sol

Volume of 3 M KCl required to cause flocculation $V_4 =$ ____ cm^3

Number of millimoles of KCl = $3\ V_4 =$ ____

Volume of sol taken = 10 cm^3

Flocculation value = Number of millimoles of KCl per dm^3 of sol

$= 3 \times 100\ V_4 =$ ____

Flocculation value of K_2SO_4 for $Fe(OH)_3$ sol

Volume of 5×10^{-3} M K_2SO_4 required to cause flocculation $V_5 =$ ____ cm^3

Number of millimoles of $K_2SO_4 = 5 \times 10^{-3}\ V_5 =$ ____

Volume of sol taken = 10 cm^3

Flocculation value = Number of millimoles of K_2SO_4 per dm^3 of sol

$= 5 \times 10^{-3} \times 100\ V_5 =$ ____

Flocculation value of $K_3[Fe(CN)_6]$ for $Fe(OH)_3$ sol

Volume of 5×10^{-4} M $K_3[Fe(CN)_6]$ required to cause flocculation $V_6 =$ ____ cm^3

Number of millimoles of $K_3[Fe(CN)_6 = 5 \times 10^{-4}\ V_6 =$ ____

Volume of sol taken = 10 cm^3

Flocculation value = Number of millimoles of $K_3[Fe(CN)_6$ per dm^3 of sol

$= 5 \times 10^{-4} \times 100\ V_6 =$ ____

Ratio of flocculation values = $V_4 : V_5 : V_6 = 1 :$ ____ $:$ ____

Result

1. The flocculation values for As_2S_3 sol are in the ratio ____ for cations of valency 1,2 and 3.
2. The flocculation values for $Fe(OH)_3$ sol are in the ratio ____ for cations of valency 1,2 and 3.

Appendix

Review Questions

1. Alum is used to purify water. Explain.
2. What do you mean by 'gold number'?

Table 7.4: CMC values of commonly used surfactants

Surfactant	*CMC (mM)*	*n*	*α*
$C_{10}N(CH_3)_3Br$	63	52	0.18
$C_{12}N(CH_3)3Br$	15	73	0.16
$C_{14}N(CH_3)_3Br$	3.6	107	0.15
$C_{14}OSO_3Na$	1.7	–	–
$C_{10}NH_3Br$	16	1100	0.07
$C_{10}(CH_3)_3Br$	18	48	0.13
$C_{10}NH_2CH_3Br$	20	670	0.07
$C_{10}NH(CH_3)_2Br$	12	69	0.08

Expected Experimental Results

Experiment 7.1 and Experiment 7.2

The Critical Micelle Concentration of sodium lauryl sulphate is approximately 6 millimol dm^{-3}. All surfactant solutions should be prepared freshly.

Experiment 7.3

For As_2S_3 sol the ratio of the flocculation values should be 52 : 0.69 : 0.093 = 1 : 0.013 : 0.0018.

For $Fe(OH)_3$ sol the ratio of the flocculation values should be 102 : 0.210 : 0.096 = 1 : 0.002 : 0.00009.

The results obtained in this experiment can vary significantly from the expected if the sol has not been purified by dialysis. The As_2S_3 sol should be dry and free of hydrochloric acid. Excess hydrogen sulphide gas should be purged from the solution by passing nitrogen gas through it.

Hints to Review Questions

1. Alum is an aluminium salt and ionises in solution to give Al^{3+} ions. Impure water often contains colloidal impurities that are negatively

charged. The Al^{3+} ions neutralise the charge on the suspended impurities and cause them to settle down, thus purifying the water.

2. Colloids can be precipitated by a number of means. To prevent their precipitation, a protective colloid is added. Zsigmondy introduced the term 'gold number' to measure the protective powers of various colloids. Gold number is the number of milligrams of the protective colloid that prevents the coagulation of 10 ml of a gold hydrosol on adding 1 ml of 10% solution of sodium chloride. For example, the gold number of gelatin is 0.005 to 0.01.

Further Exploration through the Internet

http://www.enter.net/~arsgalternatives/ea02008.htm

http://www.che.caltech.edu/research/aerosols.html

http://www.ch.kcl.ac.uk/kclchem/staff/arr/gloss.htm

http://www.bell-labs.com/org/physicalsciences/projects/colloids/colloids.html

http://www.synthashield.net/vault/colloids.html

Herman Mark

Herman Mark was born in 1895 in Austria. He served as a soldier in the Austrian army during the First World War and was awarded fourteen medals for bravery. During the fag end of the war, he was captured and held as a prisoner of war for almost a year. He spent the time learning languages and chemistry. In October 1919, he managed to escape to Vienna by bribing a prison guard.

Under the guidance of Wilhelm Schlenk at the University of Vienna, Mark worked on a doctorate thesis dealing with the synthesis and characterisation of the pentaphenylethyl free radical. In 1921, Schlenk was invited to the University of Berlin and Mark went along with him. A year later, Mark joined the Institute of Fiber Research, a new division of the Kaiser Wilhelm Institute at Berlin. Michael Polanyi was a colleague and under his direction, a broad program of X-ray crystallographic studies was initiated. Cellulose, hexamethylene tetramine, graphite, oxalic acid calomel and carbon dioxide were some of the molecules whose structure Mark investigated using X-rays.

In 1926, H. K. Meyer, a director of Germany's largest chemical corporation, the I.G. Farbenindustrie, invited Mark to become the director of a research laboratory of "high molecular" compounds in Ludwigshafern. Mark continued his studies of cellulose. In 1928 he published a pioneering paper with Meyer suggesting a polymer crystal structure that was compatible with the X-Ray diffraction pattern as well as the chemical properties. In 1930 Mark and Meyer produced the first monograph on polymer compounds, *Der Aufbau der Hochpolymeren Organischen Naturstoffe* (*The Structure of High Molecular Organic Natural Substances*). In 1932 he published a book, *Physik und Chemie der Zellulose* (*The Physics and Chemistry of Cellulose*).

In 1932, Mark was asked to leave the company because Hitler was planning to take over the German government and Mark, being a Jew, was not welcome. Mark moved back to Austria and became a professor of chemistry at the University of Vienna. There he designed the first-ever curriculum in polymer science, which included mainly the mechanism of polymerisation and the viscosity of polymer solutions.

In March 1938, Hitler's occupation of Austria forced Mark to flee to Canada where he became Research Director of the International Paper Co. Two years later Mark accepted an academic appointment at the Polytechnic Institute of Brooklyn. Polymer science in those days was very unfashionable but Mark was able to groom a number of students, like A. V. Tobolsky, P. M. Doty, B. H. Zimm, S. Krimm and R. S. Stein, who would go on to be leading lights in this area.

In 1947 the Institute of Polymer Research was set-up at the Polytechnic which offered the first ever post-graduate program of its kind in U.S.A. Mark edited the *Journal of Polymer Science*. He was one of the key people involved in setting up the polymer section of the International Union of Pure and Applied Chemistry (IUPAC). He remained active in the field, lecturing at universities and industrial laboratories, right up to his death in 1992.

8

Polymer Viscosity

8.1 Introduction

A fluid (a gas or a liquid) flowing through a tube can be thought of as consisting of a number of concentric cylindrical layers. The outermost layer is in contact with the walls of the tube and is stationary. The innermost layer at the centre of the tube flows with the maximum velocity. The layers in between have increasing velocities as one moves towards the centre of the tube. Since each layer moves at a different velocity compared to its neighbours, it behaves as if it is being dragged in a backward direction. This resistance to relative motion between the layers of the fluid is called *viscosity*.

The retarding force F created due to viscosity is proportional to the velocity gradient du/dx and to the area of contact A between the moving layers of the fluid.

$$F \propto A \frac{du}{dx}$$

$$F = \eta A \frac{du}{dx} \tag{8.1}$$

The proportionality constant η is known as the coefficient of viscosity. It may be defined as the force required to maintain a velocity difference of unity between two layers of the fluid that are unit distance apart and have unit area of contact.

Rearranging equation (8.1),

$$\eta = \frac{F}{A\, du / dx} \tag{8.2}$$

The units for η are thus kg m^{-1} s^{-1} or Pa s. In CGS units, 1 g $cm^{-1}s^{-1}$ is called 1 poise (P), after the French scientist Poiseuille. (1 Pa s = 10 poise) In 1844, Poiseuille related the viscosity of a fluid to the volume V of the liquid flowing in time t through a capillary tube of radius r, length l and pressure head P.

$$\eta = \frac{\pi r^4 t P}{8 l V} \tag{8.3}$$

This equation is known as *Poiseuille equation.*

The viscosities of some common liquids are given in Table 8.1 in Appendix 8.

8.2 Factors Affecting Viscosity

1. Temperature

The viscosity of a liquid decreases rapidly with increasing temperature. The coefficient of viscosity can be related to the absolute temperature by the Andrade equation:

$$\eta = A\, e^{(E/RT)} \tag{8.4}$$

where A is a constant and E is called the activation energy for viscous flow of the liquid.

2. Size and weight of molecules

For compounds belonging to the same chemical class, viscosity increases with increasing molecular weight.

3. Shape of molecules

Viscosity increases as the branching in the chain increases.

4. Intermolecular forces

Polar compounds are more viscous than non-polar ones because of hydrogen bonding between the molecules. For example, chloroform is heavier than water but its viscosity (0.58×10^{-3} Pa s at 293 K) is less than that of water (1×10^{-3} Pa s at 293 K).

8.3 Experimental Determination of Viscosity

The direct measurement of the viscosity of a liquid by determining the factors r, t, l, V and P in Poiseuille's equation is quite difficult and time-consuming. The method generally used is the one devised by Ostwald, which compares the viscosity of a liquid with that of another liquid of known viscosity.

The time taken by equal volumes of two liquids to flow through the same capillary is measured. Let η_1 and η_2 be the viscosities of the two liquids of densities σ_1 and σ_2 respectively. If t_1 and t_2 are their times of flow though the capillary of the viscometer then substituting in equation (8.3),

$$\eta_1 = \frac{\pi r^4 t_1 P_1}{8\, lV} \tag{8.5a}$$

and

$$\eta_2 = \frac{\pi r^4 t_2 P_2}{8\, lV} \tag{8.5b}$$

Dividing equation (8.5a) by equation (8.5b),

$$\frac{\eta_1}{\eta_2} = \frac{t_1 P_1}{t_2 P_2} \tag{8.6}$$

Since the pressure head P is directly proportional to the densities of the liquids, it follows that

$$\frac{\eta_1}{\eta_2} = \frac{t_1 \sigma_1}{t_2 \sigma_2} \tag{8.7}$$

Water is generally taken as the reference standard for viscosities. The ratio of the absolute viscosity of any liquid to that of water at the same temperature is called the *specific viscosity* of the liquid.

8.4 Viscometers

The instrument used to measure the specific viscosity of liquids is called a *viscometer*. The *Ostwald viscometer* is shown in Figure 8.1 (a).

It consists of a U-tube with two bulbs, *D* and *E*, in each of the two limbs. Bulb *D* is bigger than bulb *E*. The left limb is essentially a pipette with two definite marks *A* and *B* above and below the bulb. There is also a capillary *C* that is about 10 cm long. A definite volume of the liquid is introduced into the lower bulb using a graduated pipette. The volume of the liquid should be adjusted so that when it is forced up into the left limb up to the mark *A* by applying pressure, the meniscus in the right hand limb stands at the bottom of the bulb *E*. On releasing the liquid, the meniscus falls from *A* to *B* before the liquid stands at the same level in both the limbs. By applying a pressure to the right hand limb, the liquid is forced up into the left hand limb above the mark *A* and then allowed to fall back through the capillary under a pressure due to its own weight. A stopwatch determines the time of flow of the liquid from the mark *A* to the mark *B*.

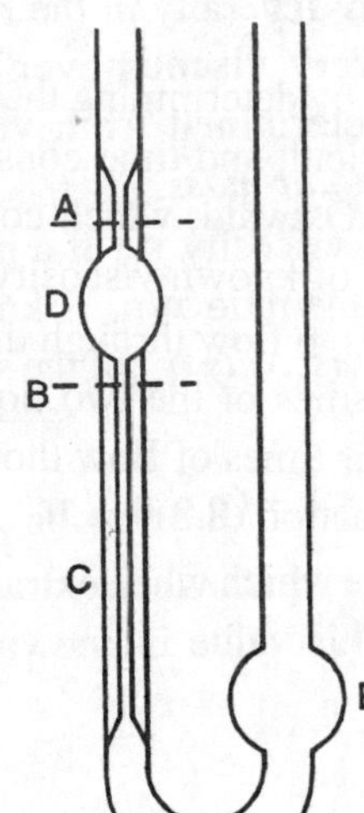

Figure 8.1(a): Ostwald Viscometer.

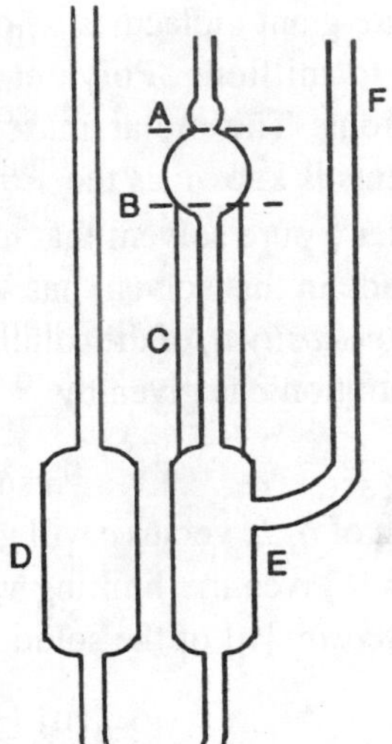

Figure 8.1(b): Ubbelhode viscometer.

A modified form of the Ostwald viscometer, known as the *suspended level viscometer* or the *Ubbelohde viscometer* is used in to study the variation of viscosity with concentration. It is shown in Figure 8.1(b). It has the advantage that the time of flow of the liquid from mark *A* to *B* is independent of the volume of the liquid introduced. Thus the solution can be diluted *in situ*. The liquid is introduced in the bulb *D*. Keeping the side arm *F* closed, the solution is forced through the capillary tube C until the meniscus stands above the mark A. On opening F while simultaneously keeping the capillary limb closed, the solution below C falls back and the column of the solution CA is held suspended. On opening the capillary limb, the solution flows through the capillary and the flow time of the meniscus from mark A to B is determined.

To change the concentration of the solution, a known volume of the solvent is introduced into D and after thorough mixing the flow time is again determined.

It is very important that the viscometer should be thoroughly cleaned and the capillary should not become obstructed by dust. If the viscometer needs cleaning, rinse it thoroughly with toluene. Then fill it with nitrating mixture. If the viscometer is wet only with water, drying is unnecessary before filling with nitrating mixture. Since there is no danger of deposition of carbon in the capillary, heat the acid on naked flame on the viscometer. This is a hazardous procedure and should be performed only by responsible laboratory staff. Following nitration, the viscometer is flushed with distilled toluene and kept in thermostat. It should be at 25.0 ± 0.1 °C and should be well stirred. The tank is best made of glass to avoid illumination problems.

8.5 Determination of Molar Mass of Polymers

Polymers are giant molecules whose molar mass is generally in the range of thousands to millions. Polymer solutions are very viscous even at low concentrations. The molar mass of polymers determined from viscosity measurements is known as the *viscosity average molar mass.*

Consider a pure solvent that has coefficient of viscosity η_0. If a polymer solution made in this solvent has viscosity η then the ratio η/η_0 is known as the *relative viscosity* η_r of the solution. The *specific viscosity* η_{sp} of the solution at a concentration c is given by

$$\eta_{sp} = \eta_r - 1 = \eta/\eta_0 - 1 = (\eta - \eta_0)/\eta_0 \qquad (8.8)$$

A graph of η_{sp}/c versus c will give a straight line which when extrapolated back to $c = 0$ gives the limiting value of η_{sp}/c. This value is known as the *intrinsic viscosity* $[\eta]$ of the solution.

$$[\eta] = \lim_{c \to 0} (\eta_{sp}/c) \qquad (8.9)$$

Intrinsic viscosity is also known as the *Staudinger index* or the *limiting viscosity number.*

It has been empirically established that the viscosity average molar mass of a polymer is related to its intrinsic viscosity by the Mark-Kuhn-Houwink-Sakurada equation

$$[\eta] \doteq K\overline{M}_v^{\,a} \qquad (8.10)$$

where K and a are constants dependent on the solvent and the polymer.

EXPERIMENTS

Experiment 8.1

Aim: To determine the limiting viscosity number (Staudinger index) of polystyrene.

Requirements

Ubbelholde viscometer, distilled toluene, 100 cm^3 standard measuring flask; 10 cm^3, 5 cm^3 and 2 cm^3 pipettes; stoppers; 250 cm^3 beaker; stop watch, etc.

Procedure

In this experiment, it is essential to minimise contamination by dust and a particularly vigorous experiment technique is accordingly imperative.

Preparation of polystyrene solution: Weigh out accurately about 1 g of polystyrene in the beaker. If the polystyrene has been dried, it will dissolve easily in the toluene but if it has not then it must be allowed to stand in 50 cm^3 toluene for 24 hours. (Cover with watch glass.)

A short contact of polymer chips with toluene is sufficient to cause the polystyrene to swell and to change its refractive index, so that it becomes virtually invisible in the solvent. When dissolution of polystyrene is complete, add the solution to a 100 cm^3 measuring flask and make up to the mark with toluene.

Use of the viscometer: The dimensions of the viscometer are fairly critical in relation to the requirements of this experiment. Rinse the viscometer with toluene. Pipette out 10 cm^3 of toluene into large bulk of the viscometer via limb A. Place it vertically in the thermostat and leave it for at least 15 minutes to come to thermal equilibrium. Remove the stopper and force the toluene up to the mark, above the level *A* in Figure 8.1(b). Note the time to the nearest 1/3rd second taken for the meniscus to pass between the marks *A* and *B*. Determine the flow time at least thrice; drain the viscometer and repeat the procedure for various polystyrene solutions. Determine the limiting viscosity.

Observations and Calculations

Flow time of toluene t_1 = ____ s

Vol. of 1% Polystyrene (cm^3)	Vol. of toluene (cm^3)	c (g cm^{-3})	Flow time t_2 (s)				η_{sp} $= (t_2 - t_1)/t_1$	η_{sp}/c (g^{-1} cm^3)
			I	II	III	mean		
20.0	0.0							
20.0	5.0							
20.0	10.0							
20.0	15.0							

Mean [η] = ______ g^{-1} cm^3

From the graph of η_{sp}/c versus c, $[\eta]$ = *y*-intercept = _____ g^{-1} cm^3.

Result

The limiting viscosity number of polystyrene

= __________(by calculations)

= __________(by graph)

Experiment 8.2

Aim: Determine the molecular mass of polystyrene from viscometry measurements.

Requirements

Ubbelhode viscometer, polystyrene, stop watch, stoppers, beakers, etc.

Procedure

Weigh out accurately about 1 g of polystyrene in the beaker. Add 50 cm^3 of and stir. Allow the solution to stand till the polystyrene dissolves completely. Transfer in a 100 cm^3 measuring flask and make up to the mark with toluene.

Take a fixed volume of pure toluene in the viscometer and measure the flow rate. Make proper dilutions of the stock solution of polystyrene and keep in the water bath at room temperature to equilibrate. Then measure the flow rate of all the dilutions and thus determine the limiting viscosity and the molecular mass of polystyrene.

Observations and calculations

Flow time of toluene t_1 = _____ s

Vol. of 1% Polystyrene (cm^3)	Vol. of toluene (cm^3)	c (g cm^{-3})	Flow time t_2 (s)				η_{sp} = $(t_2 - t_1)/t_1$	η_{sp}/c (g^{-1} cm^3)
			I	II	III	mean		
20.0	0.0							
20.0	5.0							
20.0	10.0							
20.0	15.0							

From the graph of η_{sp}/c versus c, $[\eta]$ = *y*-intercept = _____ g^{-1} cm^3

Now, $$[\eta] = 11 \times 10^{-3} M_v^{0.725}$$

Therefore, $$M_v = \{ [\eta] / 11 \times 10^{-3} \}^{1/0.725}$$

Result

Molar mass of polystyrene from viscosity measurements = __________

Experiment 8.3

Aim: *To determine the chain linkage in poly(vinyl alcohol) from viscosity measurement.*

Theory

The linear polymer poly(vinyl alcohol) (PVOH; – (–CH_2–CH_2OH–)$_n$ –) is prepared by hydrolysis of poly(vinyl acetate) (PVAc) obtained from the direct polymerisation of the monomer vinyl acetate (CH_2=CH–OOC–CH_3).

As ordinarily prepared, poly(vinyl alcohol) shows a negligible amount of branching of chains. It is somewhat unusual among synthetic high polymers that it is soluble in water.

A characteristic of interest in connection with poly(vinyl alcohol) and poly(vinyl acetate) is the consistency of orientation of the monomer unit along the chain. In the formula given above, it is assumed that all monomer units get together on head-to-tail fashion. However, occasionally a monomer unit will join into the chain in head-to-head fashion, yielding a chain of the form

reversal monomer unit

$$\text{-(-}CH_2\text{-}CHX\text{-)}_n\text{-}CH_2\text{—}\underset{\downarrow}{CHX\text{-}CHX}\text{—}CH_2\text{-}CH_2\text{-}CHX\text{-(-}CH_2\text{-}CHX\text{-)}_n\text{-}$$

head-to-head linkage

where X is the alcohol (–OH) group. The frequency of head-to-head linkage depends on the relative rates of the normal growth step reaction, α:

$$\underset{\rightarrow}{R^{\bullet}} \quad + \quad M \quad \xrightarrow{k_\alpha} \quad \underset{\rightarrow\rightarrow}{R\text{—}M}$$

and the abnormal reaction, α:

$$\underset{\rightarrow}{R^{\bullet}} \quad + \quad M \quad \xrightarrow{k_\beta} \quad \underset{\rightarrow\leftarrow}{R\text{—}M}$$

where $R^{\bullet}$ is the growing polymer radical, the arrow representing the predominant monomer orientation. The rates in turn must depend on the activation energies.

$$\frac{k_\beta}{k_\alpha} \;\alpha\; \frac{\exp\{-E_\beta / kT\}}{\exp\{-E_\alpha / kT\}}$$

$$k_\beta / k_\alpha = \exp\{-(E_\beta — E_\alpha)/kT\}$$

Presumably the activation energy for the normal reaction is the lower one in accordance with the finding of Flory and Leutner that frequency of head-to-head linkage in poly(vinyl acetate) increases with increasing polymerisation temperature.

In this experiment, the method of Flory and Leutner will be used to determine the fraction of head-to-head attachments in a single sample of poly(vinyl alcohol). The method depends on the fact that in poly(vinyl alcohol) head-to-head linkage is a 1,2-glycol structure and 1,2-glycol can be specifically and quantitatively cleaved by periodic acid or periodate ions. Treatment of poly(vinyl alcohol) with periodate should break the chain into a number of fragments, bringing about a corresponding decrease in the effective molecular weight of a specimen of poly(vinyl alcohol) before and after treatment with monodisperse periodate.

Flory and Leutner worked with monodisperse specimens of poly(vinyl alcohol) different from one another in molecular weight as determined from the osmotic pressure measurements and the intrinsic viscosity. They found that for poly(vinyl alcohol) in aqueous solution at 25°C,

$$[\eta] = 2.0 \times 10^{-4} M_n^{0.76}$$

This equation holds for a polydisperse sample of poly(vinyl alcohol) but the molecular weight in this case is M_v, the viscosity average molecular weight.

Determination of frequency of head-to-head occurrence: We wish to calculate the fraction of linkages which are head-to-head on the assumption that the degradation arises exclusively from the cleavage of 1,2-glycol structures and that all such structures are cleaved. Let us denote this ratio by Δ. It is equal to the increase in the number of molecules in the polymer divided by the total number of monomer units represented by all molecules present in the system. Since these numbers are in respective proportion to the respective molecular weight,

$$\Delta = \frac{1/M_n' - 1/M_n}{1/M_0}$$

where M_n and M_n' are number average molecular weight before and after degradation respectively, and M_0 is the monomer weight, equal to 44. Thus,

$$\Delta = 44 [1/M_n' - 1/M_n] \text{ and } \Delta = 83 [1/M_v' - 1/M_v]$$

which permits viscosity averages to be used directly.

Requirements

poly(vinyl alcohol), potassium periodate, viscometer, stop watch, etc.

Procedure

Clean the viscometer thoroughly, rinse it with distilled water and dry with acetone. Immerse it in a 25°C thermostated bath to equilibrate. Place a small flask of distilled water in the same bath to equilibrate. Then fill the viscometer with the distilled water and note the flow rate as t_1 minutes or seconds.

For the stock solution, weigh 4.0-4.5g of poly(vinyl alcohol). Transfer to a 250 cm^3 beaker. Fill three quarter full with distilled water and stir with gentle warming until the polymer is dissolved. Cool and transfer quantitatively to a 250 cm^3 standard flask, avoiding foam formation. If the solution appears contaminated, filter through Pyrex wool.

Pipette 50 cm^3 of the stock solution into 100 cm^3 volumetric flask; make up to mark with distilled water. Then again take 50 cm^3 of the above solution and dilute to 100 cm^3 in standard flask. Rinse the pipette thoroughly with distilled water and dry with acetone for every dilution.

To cleave the polymer: Pipette 50 cm^3 of the stock solution into a 250 cm^3 beaker. Add 25 cm^3 of distilled water and 0.25g of solid potassium iodate. Warm the solution to 70°C and then cool to room temperature, and make the volume to 100 cm^3 in a standard measuring flask. Place all the solutions in the water bath to equilibrate and then measure the viscosity, i.e., flow time of each solution.

Observations and calculations

Flow time of water t_1 = ______ s

For uncleaved polymer

Concentration	Flow time t_2 (s)				η_{sp}	η_{sp}/c
c (g cm^{-3})	I	II	III	mean	$= (t_2 - t_1)/t_1$	(g^{-1} cm^3)
0.016						
0.008						
0.004						

For cleaved polymer

Concentration	Flow time t_2 (s)				η_{sp}	η_{sp}/c
c (g cm^{-3})	I	II	III	mean	$= (t_2 - t_1)/t_1$	(g^{-1} cm^3)
0.008						
0.004						

From the graph of η_{sp}/c versus c for the uncleaved polymer, $[\eta]$ = ______

Now, $[\eta] = 2.0 \times 10^{-4} M_v^{0.76}$

Therefore, $M_v = \{[\eta] / 2.0 \times 10^{-4}\}^{1/0.76}$

From the graph of η_{sp}/c versus c for the cleaved polymer, $[\eta']$ = ______

Now, $[\eta'] = 2.0 \times 10^{-4} M_v'^{\,0.76}$

Therefore, $M_v' = \{[\eta'] / 2.0 \times 10^{-4}\}^{1/0.76}$

$\Delta = 83\,[1/M_v' - 1/M_v]$

Result

The fraction of head-to-head linkage in poly(vinyl alcohol) Δ = ________

Experiment 8.4

Aim: To determine the radius of a molecule by viscosity measurements. (e.g., glycerol)

Requirements

1M Glycerol solution, Ubbelhode viscometer, stop watch, distilled water.

Theory

The size of a molecule can be determined by measuring the viscosity of its solution which is related to its radius by the equation

$$\eta/\eta_0 = 1 + 2.5\Psi$$

where η is the viscosity of the solution

η_0 is the viscosity of the solvent

Ψ is the volume fraction of the solute.

The equation is based on the assumptions that the solute particles are rigid, spherical bodies without any mutual attraction. The size of the molecule must be small enough to pass through the capillary of the viscometer but much more than the size of the solvent molecule.

The volume fraction can be expressed in terms of the radius of the molecule by the equation

$$\Psi = (4/3)\,\pi\, r^3\, N_A\, C$$

where $(4/3)\,\pi\, r^3$ represent the volume of the molecule

N_A is Avogadro's constant

C is the concentration of the solution in mol dm^{-3}.

Thus a plot of η/η_0 versus concentration C will be a straight line with slope $((4/3)\,\pi\, N_A)\, r^3$.

Procedure

Prepare 0.2M, 0.4M, 0.6M and 0.8M glycerol solutions from the 1M stock solution. Clean the viscometer thoroughly and dry it. Determine the flow time t_0 for distilled water, which is the solvent in this case. Determine the flow time t for each glycerol solution.

The ratio of the viscosities η/η_0 is equivalent to the ratio t/t_0. Plot a graph of η/η_0 versus concentration c. From the slope of the graph determine the radius of the glycerol molecule.

I. *Observations and calculations*

Conc. of soln C (M) 0 (solvent)	Flow time (s)				$\eta/\eta_0 = t/t_0$
	(I)	(II)	(III)	Mean	
0.2					
0.4					
0.6					
0.8					
1.0					

From the graph of η/η_0 versus concentration C, slope = ______

$$r = \frac{[3 \times \text{slope}]^{1/3}}{[2.5 \times 4\pi\ N_A]} = \text{\_\_\_\_\_\_ nm}$$

Result

Radius of the glycerol molecule = ____ nm

Appendix

Review Questions

1. What is the effect of temperature on the viscosity of polymer solutions?
2. Define good solvent, poor solvent and theta solvent in the context of polymer solutions.
3. What is the difference between molecular weight and molar mass?
5. Does the Ubbelhode viscometer have an edge over the Ostwald viscometer?

Table 8.1: Viscosity values of some liquids

Liquid	*Temperature / (°C)*	*Viscosity / (mPa s)*
Acetic acid	25	1.155
Acetone	25	0.30
Aniline	25	3.71
Antimony, liquid	645	1.55
Benzene	20	0.65
Bismuth	285	1.61
Bromine, liquid	29	0.911
Carbon tetrachloride	20	0.97
Chloroform	20	0.58
Dimethyl sulphoxide	20	2.20
Ethanol	30	1.003
Ethyl ether	20	0.24
Ethylene glycol	20	19.9
Formic acid	20	1.804
Glucose	30	6.6×10^{13}
Glycerin	25	954
Hydrogen, liquid	–	0.011
Mercury	30	1.499
Methanol	25	0.547
Nitric acid	10	1.770
Oil, castor	30	451
Oil, light machine	38	34.2
Oil, olive	20	84.0

Oleic acid	30	25.6
Petroleum ether	20	0.30
Phenol	18	12.7
Stearic acid	70	11.6
Sucrose	109	2.8×10^6
Sulphuric acid	30	15.7
Toluene	30	0.526
Turpentine	30	1.272
Water	20	1.00

Table 8.2: *K* and *a* values for polymer-solvent systems at 313 K

Polymer	*Solvent*	$K / (10^{-3}$ *ml/g*)	*a*
Natural rubber	Benzene	18.50	0.74
Polybutadiene – high cis	Benzene	33.7	0.715
	Toluene	30.5	0.725
Polybutadiene – high trans	Toluene	29.4	0.753
Polybutadiene – cis-trans mix	Toluene	39.0	0.713
Polyethylene terephthalate	Trifluoroacetic acid	43.3	0.68
Polyisoprene – cis	Toluene	20.0	0.728
Polystyrene – atactic	Benzene	11.5	0.73
	Chloroform	4.9	0.794
	Decalin (100% trans)	61	0.53
	Dichloroethane	8.38	0.74
	Toluene	11	0.725
Polystyrene – isotactic	Benzene	10.6	0.735
	Chloroform	25.9	0.734
	Toluene	11.0	0.725
Polyvinyl acetate	Acetone	10.2	0.72
	Acetonitrile	41.5	0.62
	Benzene	56.3	0.62
	Methanol	31.4	0.60
Polyvinyl alcohol	Water	45.3	0.64
Polyvinyl chloride	Chlorobenzene	71.2	0.59
	Cyclohexane	16.3	0.77
	Tetrahydrofuran	83.3	0.83
SBR-Rubber	Cyclohexane	31.6	0.70
	Toluene	37.9	0.71

Expected Experimental Results

Experiment 8.1

The polystyrene for this experiment was obtained from Aldrich Chemical Co, Inc. The limiting viscosity number obtained was 67.0 g^{-1} cm^3.

Experiment 8.2

Using the value of limiting viscosity number obtained in Experiment 8.1, the molecular weight of polystyrene obtained was of the order of 10^5.

Experiment 8.3

The poly(vinyl alcohol) for this experiment was obtained from Spectrol, India. The fraction of head-to-head linkage in the sample was of the order of 10^{-5}.

Experiment 8.4

The glycerol for this experiment was obtained from S.D. Fine-Chem Ltd. The radius of the glycerol molecule was 2.7 nm.

Hints to Review Questions

1. As the temperature rises, the thermal energy available to the polymer molecules also rises. As they gain kinetic energy, their capacity for motion increases and hence the viscosity of the solution decreases.
2. Polymer molecules exist in solution as entangled and coiled chains, rather like snakes in a basket. The average distance between the two ends of a polymer chain is just a fraction of the distance between the two ends when the chain is fully stretched out. The interactions that exist in polymer solutions are those between segments of polymer chains, and between polymer segments and solvent molecules. In a "good" solvent, strong interaction forces between the polymer chain segments and the solvent molecules help to open up the polymer coil and aid in dissolution. In a "poor" solvent, there are no such interaction forces and the polymer chain remains tightly coiled. When a "non-solvent" is added to a polymer solution, the molecules of the non-solvent surround the polymer chain and squeeze out the solvent. The polymer is therefore no longer in solution and precipitates out. Some solvents have a characteristic temperature called *theta temperature*, above which they behave like a poor solvent and below which like a non-solvent. Such solvents are also known as *theta solvents*.
3. The *molecular mass* is the mass of a molecule of a substance. Its physical base unit is the kilogram. For atoms, the molecular mass is the *atomic mass* of an atom. The *relative molecular mass* is defined

as the ratio of the mass of a molecule of a substance to the atomic mass constant. (The *atomic mass constant* is equal to the unified atomic mass unit with symbol *amu*.) The relative molecular mass is thus a pure number. It is also called *molecular weight*. Most physical methods (osmotic pressure, static light scattering, ultracentrifugation, etc.) do not determine relative molecular masses, and thus not molecular weight but molar masses. The *molar mass* $Mi = mi/ni$ is defined as the ratio of the mass *mi* of the molecules *I* to their amount-of-substance, *ni*. It is convenient to measure the mass in grams and the amount-of-substance in moles so that molar masses and relative molecular masses become numerically identical. Note that molar masses have the physical unit mass/amount-of-substance whereas relative molecular masses (molecular weights) are dimensionless.

4. The main edge that the Ubbelhode viscometer has over the Ostwald viscometer is that dilution of the polymer solution can be made inside the bulb of the viscometer itself. The experiment is started with a single solution of known concentration. Subsequent dilutions are made by adding known volume of the pure solvent and mixing inside the viscometer. The Ostwald viscometer has to be emptied cleared and refilled every time a new dilution is required.

Further Exploration through the Internet

http://www.cheresources.com/polyvisczz.shtml
http://www.soest.hawaii.edu/SPACEGRANT/class_acts/ViscosityTe.html
http://www.umr.edu/~wlf/MW/viscos.html
http://physics.bu.edu/py105/notes/Viscosity.html
http://www.processassociates.com/process/convert/cf_vab.htm
http://xtronics.com/viscosity.htm

Michael Faraday

Michael Faraday was born in 1791 in London, England. At the age of fourteen he became an apprentice to a bookbinder. During his seven-year apprenticeship he developed an interest in science, and in particular, chemistry. Towards the end of his apprenticeship, in 1812, he attended Sir Humphry Davy's lectures at the Royal Institution. Faraday took notes during the lectures and presented them later to Davy, asking for a job. In 1813 he was granted his desire and was taken on as a laboratory assistant at the Royal Institution. There he immersed himself in the study of chemistry, becoming a skilled analytical chemist. In 1823 he discovered that chlorine could be liquefied and in 1825 he discovered benzene.

However, Faraday's greatest achievements were concerned with electricity. In 1820 the Danish scientist, Oersted, discovered the phenomenon of electromagnetism. The following year was one of the most important in his life. In his basement laboratory at the Royal Institution, Faraday conducted a series of experiments that culminated in the discovery of rotation caused by an electromagnetic field – the principle behind the electric motor. 1821 was also the year in which he married and was promoted to the post of the Superintendent of the House.

In1831 he discovered electromagnetic induction, the principle behind the transformer and the generator. This discovery propelled electricity from the status of a mere scientific curiosity to a powerful technology. Between 1832 and 1834 he enunciated the laws of electrolysis, which today bear his name. Working with William Whewell he coined the terms associated with this field – electrode, electrolyte, anode, cathode, etc. Faraday then worked on a new theory of static electricity and electrical induction, which negated the old ideas that electricity was some sort of a fluid.

Faraday experimented with magnets and found that not just matter, even light was affected by magnetism. He proposed that magnetism is an inherent property of matter. His work laid the basis for the field theory of magnetism, which was mathematised by William Thomson (later Lord Kelvin) and then by James Clerk Maxwell.

During the later years of his life, Faraday suffered from ill health. His scientific activity declined significantly, and in1867 he passed away.

9

Conductance of Solutions

9.1 Introduction

Life without electricity is hard to imagine. Electricity plays an important role in chemistry as well. Many chemical reactions generate electricity and they are harnessed to make batteries and fuel cells. The passage of electricity through certain substances causes chemical reactions.

A substance that allows an electric current to flow is called a *conductor* while one that does not is called an *insulator*. Conductors may be metals or electrolytes. The flow of mobile, negatively charged electrons through a metallic conductor generates electric current. *Electrolytes* are substances, which in the molten state or in solution conduct electricity by the movement of ions. The passage of electric current through an electrolyte brings about chemical changes, which may include decomposition and combination. This phenomenon is called *electrolysis*.

Electrolytes can be classified as weak or strong depending on the amount of electric current they can conduct. *Weak electrolytes* ionise partially in the molten state or in solution, which makes them low conducting. *Strong electrolytes* ionise completely, which enables them to carry a large amount of electric current.

9.2 Electrolysis

In electrolysis, metal or graphite electrodes are dipped into the electrolyte, which may be in a molten state or in solution. The electrodes are connected to the terminals of a battery by means of conducting wires (Figure 9.1).

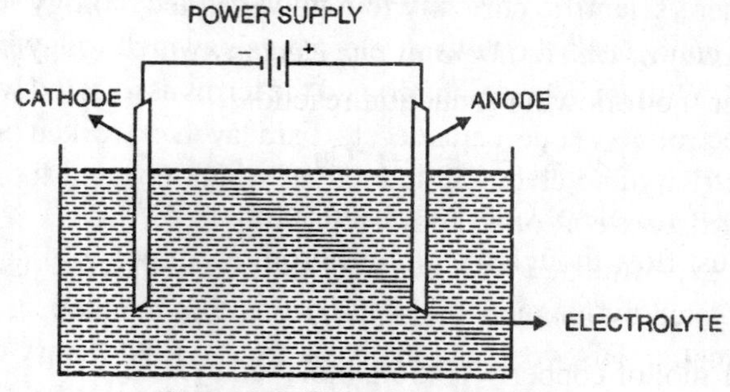

Figure 9.1: Electrolysis apparatus.

The *anode* is the electrode connected to the positive terminal of the battery. Current enters the electrolyte through the anode. Being positively charged, the anode attracts anions towards it. The discharge of anions at the anode is accompanied by a loss of electrons. This is called *oxidation*. The *cathode* is the electrode connected to the negative terminal of the battery. Current leaves the electrolyte through the cathode. Cations migrate towards the negatively charged electrode and are discharged there by taking up electrons. Hence *reduction* takes place at the cathode. Thus the reactions taking place during electrolysis are always oxidation-reduction or *redox* reactions.

In 1834 Michael Faraday quantified the phenomenon of electrolysis by enunciating two laws that are now called *Faraday's Laws of Electrolysis.* Faraday's Laws are the most exact laws in physical chemistry because they hold at all temperatures, in all concentrations, at all pressures and in different solvents.

9.2.1 First law of electrolysis

The weight W of an ion discharged during electrolysis is proportional to the quantity of electricity Q passed. If I amperes of current are passed through the electrolyte for time t seconds then the amount of electric charge Q is given by

$$Q = I\,t \tag{9.1}$$

According to the statement of the First Law,

$$W \propto Q$$

$$W = zQ \tag{9.2}$$

where z is the proportionality constant and is called the electrochemical equivalent of the ion. When a current strength of 1 ampere flows through the electrolyte for 1 second, the weight of the ion discharged is equal to its electrochemical equivalent.

The charge on one mole of electrons is called one faraday, F. If N_A is the Avogadro constant and e is the charge on one electron then

$$\begin{aligned} F &= N_A\, e \qquad (9.3)\\ &= (6.023 \times 10^{23}\ \text{mol}^{-1})(1.6 \times 10^{-19}\ \text{C})\\ &= 96,485\ \text{C mol}^{-1} \end{aligned}$$

One faraday of charge liberates one electrochemical equivalent of an ion.

Consider the following reduction reaction:

$$Cu^{2+}_{(aq)} + 2e^- \rightarrow Cu_{(s)} \tag{9.4}$$

To deposit 1 mole of Cu^{2+} ions from solution, 2 moles of electrons (i.e., 2 faradays) must flow though the circuit. From equation (9.2),

$$z = W/Q$$

If $W = 1$ mol of copper = 63.56 g mol^{-1} and $Q = 2\,F$,

$$z = 63.56/2 = 31.78\ \text{g C}^{-1}$$

Hence the electrochemical equivalent of copper is 31.78 g C^{-1}.

In general, to deposit one mole of the metal *M* from a solution containing the ion M^{z+}, *z* faradays of charge are required, and the electrochemical equivalent of the metal is its atomic weight divided by *z*.

9.2.2 Second law of electrolysis

When the same quantity of electricity passes though different electrolytes arranged in series, the weights of the different ions liberated at the electrodes are in the ratio of their electrochemical equivalents.

For example, if two solutions of $CuSO_4$ and $AgNO_3$ are arranged in series, and the same quantity of electricity is passed through them then

$$\frac{\text{Wt of Cu deposited}}{\text{Wt of Ag deposited}} = \frac{\text{electrochemical equivalent of Cu}}{\text{electrochemical equivalent of Ag}} \quad (9.5)$$

This law can be used to keep track of how much current has passed through the circuit during the electrolysis by employing a device known as a *coulometer*. A number of coulometers have been devised. The most commonly used one is the silver coulometer. The silver coulometer is a platinum crucible containing 15% $AgNO_3$ solution in which an electrode of pure silver is dipped. The platinum crucible acts as the cathode and the silver electrode as the anode. During electrolysis silver gets deposited in the platinum crucible. The difference in the mass of the crucible before and after electrolysis gives the mass of the silver deposited. The quantity of electricity *Q* that passed through the circuit is given by

$$Q = z F m/M \quad (9.6)$$

where *z* is the charge on the silver ion ($z = 1$ for Ag^+)

F is the Faraday constant

m is the mass of silver metal deposited in g

M is the molar mass of silver ($M = 107.88$ g mol^{-1} for Ag)

9.3 Conductance of Solutions

Like metallic conductors, electrolytes obey Ohm's Law. The current *I* flowing through an electrolyte is directly proportional to the potential difference *V* applied across the electrodes dipping in it.

$$V \propto I$$

$$V = IR \quad (9.7)$$

The proportionality constant *R* is known as the resistance of the electrolyte. Since we are concerned with the conducting power of the electrolyte, we take the reciprocal of the resistance.

Thus conductance, *G*, is given by

$$G = 1/R \quad (9.8)$$

The SI unit of conductance is *siemens* (S).

The resistance of a uniform conductor is directly proportional to its length l, and inversely proportional to its area of cross-section A.

$$R \propto l/A$$

$$R = \sigma\, l/A \tag{9.9}$$

where the proportionality constant σ is called *resistivity*.

Taking reciprocal of equation (9.9),

$$1/R = 1/\sigma \,.\, A/l \tag{9.10}$$

The reciprocal of resistivity is called the *conductivity* χ *(kappa)*.

Substituting $G = 1/R$ and $\chi = 1/\sigma$ in equation (9.10),

$$G = \chi \,.\, A/l \tag{9.11}$$

or

$$\chi = G \,.\, l/A \tag{9.12}$$

Thus the conductivity of an electrolyte is the conductivity of a solution taken in a cell in which the electrodes are a unit distance apart and have unit cross-section area. In other words, it is the conductivity of unit volume of the solution. If the cross-section area of the electrodes is 1 m^2 and the distance between the electrodes is 1 m then the SI unit of conductivity is S m^{-1}.

9.3.1 Variation of conductivity with concentration

The ions present in an electrolyte are responsible for carrying electric current. The number of ions present in a given volume of the solution depends on the concentration of the solution. Therefore, the conductivity of a solution varies with its concentration.

9.3.1.1 Molar conductivity Λ_m

The molar conductivity of a solution at a specified concentration and temperature is the conductivity of that volume of the solution which contains one mole of the solute and is placed between two electrodes unit distance apart and having sufficient cross-section area to hold the whole of the solution.

If one mole of an electrolyte is dissolved in V cm^3 of the solution then the solution will cover V cm^2 of the electrodes placed 1 cm apart. Substituting $G = \Lambda_m$, $l = 1$ cm and $A = V$ cm^2 in equation (9.11),

$$\Lambda_m = \chi\, V \tag{9.13}$$

If the concentration of the solution is c M then 1000 cm^3 of the solution contain c moles of the electrolyte. Thus the volume V that contains one mole of the electrolyte is given by

$$V = 1000/c \tag{9.14}$$

Substituting equation (9.14) in equation (9.13),

$$\Lambda_m = 1000\, \chi / c \tag{9.15}$$

Molar conductivity has units S m^2 mol^{-1} though S cm^2 mol^{-1} is more commonly used.

9.3.1.2 Equivalent conductivity Λ

The equivalent conductivity of a solution at a specified concentration and temperature is the conductivity of that volume of the solution which contains one gram equivalent of the solute and is placed between two parallel electrodes unit distance apart and large enough to contain the whole of the solution between them.

If 1 gram equivalent of the electrolyte is dissolved in V cm^3 of the solution then the solution will cover V cm^2 of the electrodes placed 1 cm apart. Substituting $G = \Lambda$, $l = 1$ cm and $A = V$ cm^3 in equation (9.11).

$$\Lambda = \chi V \tag{9.16}$$

If the concentration of the solution is c N then 1000 cm^3 of the solution contains c gram equivalent of the electrolyte. Thus the volume V that contains one gram equivalent of the electrolyte is given by $1000/c$. Substituting in equation (9.16),

$$\Lambda = 1000\,\chi/c \tag{9.17}$$

Λ has units S m^2 eq^{-1} though S cm^2 eq^{-1} is more commonly used.

9.3.2 Relationship between equivalent conductivity and molar conductivity

An electrolyte on dissociation in solution furnishes cations and anions. Let z be the total positive charge carried by the cation (or the total negative charge carried by the anions) formed on dissociation of one mole of the electrolyte.

For example, aluminium sulphate undergoes dissociation as follows:

$$Al_2(SO_4)_3 \longleftrightarrow 2Al^{3+} + 3(SO_4)^{2-}$$

Each aluminium ion carries a +3 charge. Since two aluminium ions are formed per aluminium sulphate molecule, the total positive charge is +6. Each sulphate ion carries a –2 charge. Since two sulphate ions are formed per aluminium sulphate molecule, the total negative charge is –6. Hence $z = 6$.

The equivalent conductivity of the electrolyte is its molar conductivity divided by z.

$$\Lambda = \Lambda_m/z \tag{9.18}$$

9.3.3 Weak and strong electrolytes

If all electrolytes behaved ideally, the molar conductivity of a completely dissociated salt would be strictly proportional to the number of ions present in unit volume and therefore molar conductivity would be independent of concentration.

In reality, the molar conductivity of electrolytes increases with dilution. In the case of strong electrolytes, this is due to the fact that electrostatic interactions between oppositely charged ions decreases as the concentration of the solution decreases. The molar conductivity of a strong electrolyte at infinite dilution Λ^∞_m can be approximated by extrapolating to zero concentration the curve of its molar conductivity versus concentration.

In the case of weak electrolytes the major reason for the increase in molar conductivity at increasing dilution is the fact that the degree of dissociation of the electrolyte increases with decreasing concentration. The molar conductivity at infinite dilution is difficult to determine using the extrapolation method because the rapid increase in molar conductivity near zero concentration makes extrapolation difficult and might result in gross errors. Instead, two alternative methods can be applied: one using Kohlrausch's law of independent migration of ions, and the other using Ostwald's Dilution Law.

9.3.3.1 Kohlrausch's law of independent migration of ions

In 1874 the German chemist Kohlrausch showed that Λ^∞_m of an electrolyte is the sum of the contributions from its individual ions. If the molar conductivities of the cations and anions at infinite dilution are denoted by λ^∞_+ and λ^∞_- respectively then

$$\Lambda^\infty_m = \nu_+\lambda^\infty_+ + \nu_-\lambda^\infty_- \tag{9.19}$$

where ν_+ and ν_- are the numbers of cations and anions per formula unit of electrolyte and λ^∞_+ and λ^∞_- are called *ionic molar conductivities at infinite dilution*.

For example, the molar conductivity of acetic acid (a weak electrolyte) at infinite dilution can be expressed as

$$\Lambda^\infty_{HAc} = \nu^\infty_{H+} + \nu^\infty_{Ac-}$$

since $\nu_+ = \nu_- = 1$.

To each side of this equation, the Λ^∞_m of the salt of a strong acid and strong base, such as NaCl, is added.

$$\begin{aligned}\Lambda^\infty_{HAc} + \Lambda^\infty_{NaCl} &= \lambda^\infty_{H+} + \lambda^\infty_{Ac-} + \Lambda^\infty_{NaCl}\\ &= \lambda^\infty_{H+} + \lambda^\infty_{Ac-} + \lambda^\infty_{Na+} + \lambda^\infty_{Cl-}\\ &= (\lambda^\infty_{H+} + \lambda^\infty_{Cl-}) + (\lambda^\infty_{Na+} + \lambda^\infty_{Ac-})\\ &= \Lambda^\infty_{HCl} + \Lambda^\infty_{NaAc}\end{aligned}$$

Therefore,

$$\Lambda^\infty_{HAc} = \Lambda^\infty_{HCl} + \Lambda^\infty_{NaAc} - \Lambda^\infty_{NaCl} \tag{9.20}$$

The molar conductivities on the right hand side of this equation are those of strong electrolytes and hence can be determined by the extrapolation of a Λ_m versus c graph.

The values of ionic molar conductivities for individual ions are given in Table 9.1 in Appendix 9.

9.3.3.2 Ostwald's dilution law

Weak electrolytes do not dissociate completely in solution. There exists an equilibrium between the undissociated electrolyte and its ions. Consider the dissociation of acetic acid in water:

$$CH_3COOH + H_2O \longleftrightarrow CH_3COO^- + H_3O^+$$

The equilibrium constant for this reaction is

$$K = \frac{[H_3O^+][CH_3COO^-]}{[CH_3COOH]} \tag{9.21}$$

As the concentration of acetic acid in the aqueous solution decreases the equilibrium shifts further towards the right. At infinite dilution, it is assumed that acetic acid is completely dissociated. Since the conductivity of the solution depends on the number of ions in solution, the greater the degree of dissociation of acetic acid the greater will be its conductivity.

If α is the degree of dissociation and c the molar concentration of acetic acid in solution then at equilibrium,

$$[H_3O^+] = [CH_3COO^-] = \alpha c \quad [CH_3COOH] = (1 - \alpha)c \tag{9.22}$$

Ignoring activity coefficients, K is given by

$$K = \frac{\alpha^2 c}{1 - \alpha} \tag{9.23}$$

If the molar conductivity of acetic acid at infinite dilution is Λ^∞_m then its molar conductivity at concentration c will be

$$\Lambda_m = \alpha \Lambda^\infty_m \tag{9.24}$$

Substituting equation (9.24) in equation (9.23),

$$K = \frac{c\Lambda_m^2}{\Lambda^\infty_m (\Lambda^\infty_m - \Lambda_m)} \tag{9.25}$$

Rearranging this equation,

$$\frac{1}{\Lambda_m} = \frac{1}{\Lambda^\infty_m} + \frac{c\Lambda_m}{K(\Lambda^\infty_m)^2} \tag{9.26}$$

Equation (9.26) is known as *Ostwald's Dilution Law.*

A graph of $1/\Lambda_m$ versus $c\Lambda_m$ will give a straight line with slope $1/K(\Lambda^\infty_m)^2$ and intercept $1/\Lambda^\infty_m$. From the values of the slope and the intercept, the values of K and Λ^∞_m can be determined.

9.4 The Drift Speed and Mobility of Ions

When a potential difference is applied across two electrodes dipped in an electrolyte, the ions in the solution accelerate towards the electrodes. As they move through the solvent, they experience a retarding force due to friction with solvent molecules. When the accelerating force matches the retarding force, the ions reach a terminal speed called *drift speed.* It decides the rate at which charge is transported.

The drift speed is proportional to the strength of the applied electric field. The proportionality constant is called the *mobility* of the ion, defined as the drift speed acquired in an electric field of unit strength. It is related to the molar conductivity as follows:

$$\lambda = z u F \tag{9.27}$$

where λ is the molar conductivity of the ion
z is the charge on the ion
u is the mobility of the ion
F is the Faraday constant.

The mobility of a number of ions in water at 298 K is given in Table 9.2 in Appendix 9.

The molar conductivity of the electrolyte is given by the sum of the molar conductivities of its individual ions.

$$\Lambda_m = (\nu_+ z_+ u_+ + \nu_- z_- u_-)F \tag{9.28}$$

9.4.1 Transport numbers

The *transport number* of an ion is the fraction of the total current carried by that ion.

If t_+ and t_- are the transport numbers of the cations and anions respectively of an electrolyte,

$$t_+ = \nu_+ \lambda_+ /(\nu_+ \lambda_+ + \nu_- \lambda_-) \quad t_- = \nu_- \lambda_- /(\nu_+ \lambda_+ + \nu_- \lambda_-) \tag{9.29}$$

It follows from the above equation that the sum of the transport numbers of all the ions in the electrolyte must equal unity.

The transport number is related to the mobility of the ion as follows:

$$t_+ = u_+ /(u_+ + u_-) \qquad t_- = u_- /(u_+ + u_-) \tag{9.30}$$

The transport number of an ion in solution depends on the other ions present in solution along with it as well as on their relative concentrations. It can be experimentally determined by either Hittorf's method or the moving boundary method.

9.4.2 Hittorf's method for determination of transport numbers

In Hittorf's method, the electrolytic cell consists of three compartments: the cathode compartment, the middle compartment and the anode compartment (Figure 9.2). When a direct current I is passed through the electrolyte for a time interval Δt, the total charge Q passed is $I\Delta t$. This causes $(1/z_+)(Q/F)$

equivalents of the cation to be formed in the anode compartment. These cations are attracted towards the negatively charged cathode and they migrate out of the anode compartment. During this process of migration, the cations transport a fraction t_+ of the total charge. Thus the net change in the equivalents of the cation in the anode compartment is given by

$$\text{Net change} = (1 - t_+)(1/z_+)(Q/F)$$
$$= (t_- / z_+)(Q/F) \qquad (9.31)$$

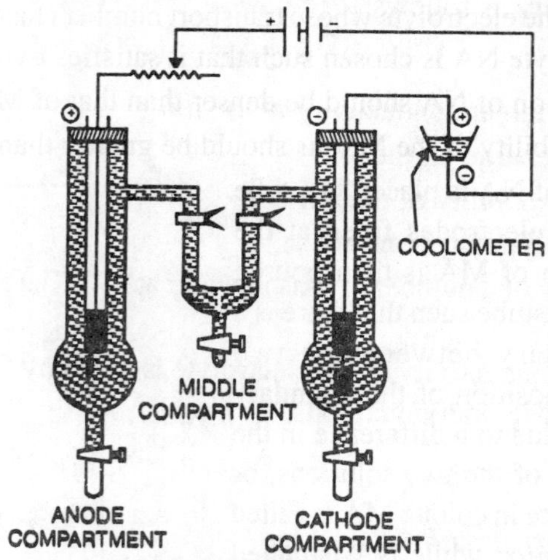

Figure 9.2: Hittorf cell.

Table 9.3 shows the net change in the number of moles of cations and anions in each of the compartments.

Table 9.3: The net change in the number of moles of cations and anions in each of the compartments of the Hittorf cell after electrolysis

Ion	*Net change in number of moles*		
	Cathode compartment	*Middle compartment*	*Anode compartment*
Cation	$(t_+ - 1)/z_+ = -t_-/z_+$	0	$(1 - t_+)/z_+ = t_-/z_+$
Anion	$-t_-/z_-$	0	t_-/z_-

Many practical difficulties arise when determining transport numbers using Hittorf's method. The change in the concentration of the ions in the different compartments leads to the formation of a concentration gradient. This results in the diffusion of ions from regions of higher concentration to those of lower concentration. It also results in density differences. Diffusion can be minimised by curtailing the duration of the experiment. The experiment should not be cut short too much because a small amount of charge will produce small concentration changes, which will be difficult to determine accurately. If large currents are passed, the electrolyte heats up which causes convection. In concentrated electrolyte solutions, there is a significant concentration of ion-pairs that carry a part of the current. In spite of all these difficulties, the Hittorf's method gives reasonably good values of transport numbers.

9.4.3 Moving boundary method for determination of transport numbers

The moving boundary method, in contrast with Hittorf's method, gives very accurate values of transport numbers.

Let MA be the electrolyte whose transport numbers have to be determined. Another electrolyte NA is chosen such that it satisfies two criteria:

- A solution of NA should be denser than that of MA.
- The mobility of the M ions should be greater than that of the N ions.

A solution of NA is placed in a tube containing two electrodes fixed at the ends. A solution of MA is then poured carefully into the tube such that there is a distinct boundary between the two solutions. The position of the boundary may be visible due to a difference in the refractive index of the two solutions, or due to a difference in colour. MA is called the *leading solution* while NA is called the *indicator solution.*

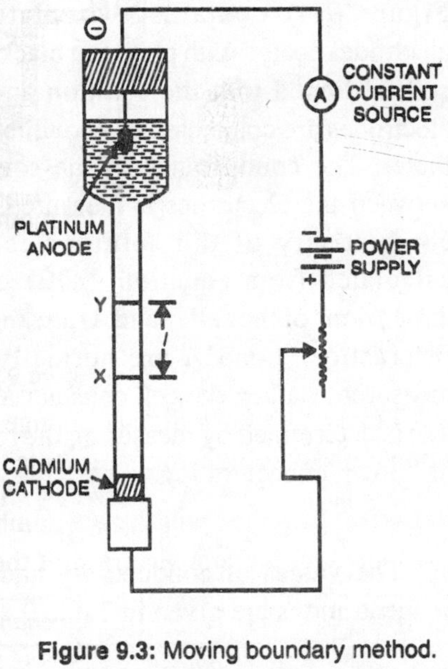

Figure 9.3: Moving boundary method.

Let the boundary between the two solutions be at position X as shown in Figure 9.3. Let a direct current of strength I be passed through the solution for a time interval Δt. The new position of the boundary is now Y. If l is the length between X and Y and A is the cross-sectional area of the tube then the volume travelled by M ions is lA. If c is the concentration of M ions in mol dm^{-3} then the number of ions that have travelled the length is XY is $cN_A lF$. If the charge on each M ion is z_+ then the charge carried by the M ions is $z_+ cN_A lF$. Since the total charge passed through the solution is $I\Delta t$, the transport number of M ions is

$$t_+ = \frac{z_+ cN_A lF}{I\Delta t} \tag{9.32}$$

The accuracy of the results obtained by the moving boundary method depends very much on maintaining a sharp boundary between the leading solution and the indicator solution. The concentration of the indicator solution should be less than that of the leading solution. Convection currents and heating of the solution should be avoided by keeping low the magnitude of the current passed. The tube used should be narrow and can be surrounded by water to maintain a constant temperature. The electrode reactions should not liberate any gases at the voltage used in the experiment.

9.5 Experimental Determination of Conductivity

9.5.1 The conductivity cell

The conductance of a solution cannot be measured by passing a direct current through the solution because electrolysis causes changes in the concentrations of the ions. To avoid electrolysis, an alternating current is used. A conductivity cell (see Figure 9.4) contains two platinum electrodes coated with platinum black. The cell is dipped into the solution and the electrodes are connected to a conductivity meter. The conductance of the solution between the electrodes is measured. The conductivity of the solution can be calculated from equation (9.12) if the dimensions of the cell (l and A) are known. In practice, l and A are normally not measured. Rather, the cell constant K_{cell} of the conductivity meter (defined as l/A) is determined by measuring the resistance of a potassium chloride solution of known concentration.

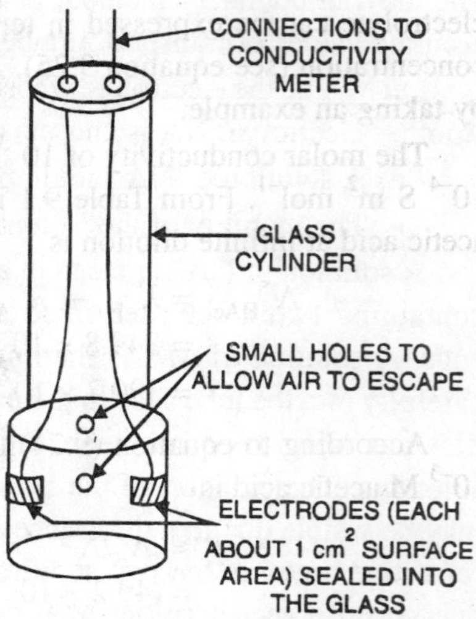

Figure 9.4: Conductivity cell.

$$K_{cell} = \chi R \tag{9.33}$$

The values of conductivity and molar conductivity of KCl at various concentrations are given in Table 9.4 in Appendix 9.

9.5.2 Temperature control

Temperature control is important in conductance measurements. The use of a thermostat gives results with an accuracy of ± 0.1%. For more accurate measurements an oil bath should be used. In the case of conductometric titrations, a constant temperature can be maintained by surrounding the titration vessel with a large water bath. Most modern conductivity meters incorporate an automatic temperature compensator, which is itself immersed in the solution along with the conductivity cell.

9.5.3 Solvent correction

In exact conductivity measurements, solvent correction must be applied. The solvent should be purified so that its conductivity is negligible. The conductivity of ordinary distilled water is of the order of 10^{-6} S cm^{-1} at 25°C, the chief impurities being dissolved gases like carbon dioxide and ammonia. The conductivity of the solvent is subtracted from the conductivity of the solution to get the conductivity of the electrolyte.

9.6 Experimental Determination of Dissociation Constants

According to Ostwald's Dilution Law, the dissociation constant of a weak electrolyte can be expressed in terms of its molar conductivity at a certain concentration (see equation 9.25). The use of this equation can be illustrated by taking an example.

The molar conductivity of 10^{-3} M acetic acid in water at 298 K is 49.2 × 10^{-4} S m^2 mol^{-1}. From Table 9.1 in Appendix 9, the molar conductivity of acetic acid at infinite dilution is

$$\begin{aligned}\Lambda^{\infty}_{HAc} &= \lambda^{\infty}_{H+} + \lambda^{\infty}_{Ac-}\\ &= 349.8 \times 10^{-4}\ \text{S m}^2\ \text{mol}^{-1} + 40.9 \times 10^{-4}\ \text{S m}^2\ \text{mol}^{-1}\\ &= 390.7 \times 10^{-4}\ \text{S m}^2\ \text{mol}^{-1}\end{aligned}$$

According to equation (9.24) the approximate degree of dissociation of 10^{-3} M acetic acid is

$$\begin{aligned}\alpha &\approx \Lambda_m/\Lambda^{\infty}_m\\ &= (49.2 \times 10^{-4}\ \text{S m}^2\ \text{mol}^{-1})/(390.7 \times 10^{-4}\ \text{S m}^2\ \text{mol}^{-1})\\ &= 0.1259\end{aligned}$$

$$[H_3O^+] = [CH_3COO^-] = \alpha c = 0.1259 \times 10^{-3}\ \text{M} = 1.259 \times 10^{-4}\ \text{M}$$

$$[CH_3COOH] = (1 - \alpha)\, c = (1 - 0.1259) \times 10^{-3}\ \text{M} = 8.741 \times 10^{-4}\ \text{M}$$

The approximate value of the dissociation constant is thus:

$$K' = \frac{[H_3O^+][CH_3COO^-]}{[CH_3COOH]}$$

$$= \frac{(1.259 \times 10^{-4})^2}{8.741 \times 10^{-4}} = 1.813 \times 10^{-5}$$

When the values of K' at different concentrations of acetic acid are plotted against the concentration of acetic acid, a straight line is obtained for values of concentration less than 2 × 10^{-4} M. When this line is extrapolated to zero concentration, the value of the true thermodynamic dissociation constant is obtained.

Considering that the thermodynamic dissociation constant of acetic acid is 1.749 × 10^{-5}, the approximate value obtained above (1.813 × 10^{-5}) is quite good. Hence, for most practical purposes the approximate equation (9.25) is used.

9.7 Experimental Determination of Ionic Product of Water

Water molecules undergo very weak dissociation.

$$H_2O \longleftrightarrow H^+ + OH^-$$

The equilibrium constant of this reaction is

$$K_{eq} = \frac{[H^+][OH^-]}{[H_2O]} \tag{9.34}$$

Since the concentration of water doesn't change appreciably due to dissociation, $K_{eq}[H_2O]$ is almost constant. This constant is known as the ionic product of water K_w.

$$K_w = [H^+][OH^-] \tag{9.35}$$

Since the ions, H^+ and OH^-, are formed in very small quantities, their concentrations can be considered to be equal to their activities.

$$K_w = C_{H+} \cdot C_{OH-} \tag{9.36}$$

The units of K_w here are (concentration)^2. To make K_w dimensionless, each concentration term is divided by the standard concentration C^0, which is defined as 1 mol dm^{-3}.

$$K_w = \frac{C_{H+}}{C^0} \cdot \frac{C_{OH-}}{C^0} \tag{9.37}$$

In pure water,

$$C_{H+} = C_{OH-} = C^0(K_w)^{1/2} \tag{9.38}$$

The conductivity χ of pure water is given by

$$\chi = C_{H+}\lambda_{H+} + C_{OH-}\lambda_{OH-} \tag{9.39}$$

From equation (9.38), this is the same as

$$\chi = C^0(K_w)^{1/2}(\lambda_{H+} + \lambda_{OH-}) \tag{9.40}$$

Since the concentrations of the H^+ and OH^- ions are very small, their ionic conductivities may be taken as equal to their ionic conductivities at infinite dilution. From Table 9.1 in Appendix 9,

$$\lambda^\infty_{H+} + \lambda^\infty_{OH-} = 548.1 \times 10^{-4}\ \text{S m}^2\ \text{mol}^{-1}$$

Kohlrausch obtained a value 5.5×10^{-6} S m^{-1} for the conductivity of pure water at 25°C. Substituting these values in equation (9.40),

$$K_w = \frac{\chi^2}{(C^0)^2(\lambda_{H+} + \lambda_{OH-})^2} \tag{9.41}$$

$$= \frac{(5.5 \times 10^{-6}\ \text{S m}^{-1})^2}{(1\ \text{mol dm}^{-3} \times 10^3\ \text{m}^{-3}/\text{dm}^{-3})^2(548.1 \times 10^{-4}\ \text{S m}^2\ \text{mol}^{-1})^2}$$

$$= 1.01 \times 10^{-14}.$$

9.8 Experimental Determination of Hydrolysis Constants

A salt formed from a weak acid or weak base will be partially or completely decomposed by interaction with water. This phenomenon is known as hydrolysis. For example, aniline hydrochloride salt is formed from a weak base (aniline) and a strong acid (hydrochloric acid). Its reaction with water is

$$C_6H_5NH_3{}^+Cl^- + H_2O \longleftrightarrow C_6H_5NH_2 + H_3O^+ + Cl^-$$

Omitting the Cl^- ion which is common on both sides,

$$C_6H_5NH_3{}^+ + H_2O \longleftrightarrow C_6H_5NH_2 + H_3O^+$$

The equilibrium constant of this reaction is

$$K_{eq} = \frac{[C_6H_5NH_2][H_3O^+]}{[C_6H_5NH_3^+Cl^-][H_2O]} \tag{9.42}$$

Since the concentration of water does not change appreciably, $K_{eq}[H_2O]$ is almost constant and is known as the *hydrolysis constant* K_h.

$$K_h = \frac{[C_6H_5NH_2][H_3O^+]}{[C_6H_5NH_3^+]} \tag{9.43}$$

The fraction of the salt that is dissociated into ions is called the degree of hydrolysis.

If α is the degree of hydrolysis and c mol dm^{-3} the concentration of ammonium chloride solution,

$$[C_6H_5NH_2] = [H^+] = \alpha c \qquad [C_6H_5NH_3{}^+] = (1 - \alpha)c \tag{9.44}$$

Substituting these values in (9.43)

$$K_h = \frac{\alpha^2 c}{1 - \alpha} \tag{9.45}$$

In the experimental procedure to determine the hydrolysis constant from conductivity measurements the molar conductivity Λ_m of the aqueous solution of aniline hydrochloride at a certain concentration c is measured.

The molar conductivity Λ_m' of aniline hydrochloride at the same concentration in presence of excess of the weak base, aniline, is also measured. In presence of excess of aniline, the hydrolysis of aniline hydrochloride is suppressed and hence the molar conductivity of unhydrolysed aniline hydrochloride is obtained.

The molar conductivity Λ_{HCl} of the strong acid HCl is also measured. Alternatively, it can be assumed that the strong acid is completely dissociated in solution. Its molar conductivity can then be calculated as the sum of the ionic molar conductivities at infinite dilution of its constituent ions.

The molar conductivity of the aniline hydrochloride solution can be expressed as the sum of the molar conductivity due to unhydrolysed aniline hydrochloride and that of the hydrochloric acid produced.

If α is the degree of hydrolysis,

$$\Lambda_m = (1-\alpha)\,\Lambda_m' + a\,\Lambda_{HCl} \tag{9.46}$$

This equation can be rearranged to give

$$\alpha = \frac{\Lambda_m - \Lambda'_m}{\Lambda_{HCl} - \Lambda'_m} \tag{9.47}$$

Equation (9.45) can then be used to determine the hydrolysis constant of the salt.

Consider an example. The conductivity of 0.001 M aqueous solution of aniline hydrochloride was found to be 1.50 × 10^{-2} S m^{-1} at 298 K. In the presence of excess of aniline the value came down to 1 × 10^{-2} S m^{-1}.

Molar conductivity of the hydrolysed salt is

$$\Lambda_m = \frac{1.50\times10^{-2}\ \text{S m}^{-1}}{(0.001\ \text{mol dm}^{-3})\times(10^{3}\ \text{m}^{-3}/\text{dm}^{-3})}$$
$$= 1.50\times10^{-2}\ \text{S m}^2\ \text{mol}^{-1}$$

Molar conductivity of the unhydrolysed salt is

$$\Lambda_m' = \frac{1\times10^{-2}\ \text{S m}^{-1}}{(0.001\ \text{mol dm}^{-3})\times(10^{3}\ \text{m}^{-3}/\text{dm}^{-3})}$$
$$= 1.00\times10^{-2}\ \text{S m}^2\ \text{mol}^{-1}$$

From Table 9.1 in Appendix 9,

$$\Lambda_{HCl} = \lambda^{\infty}{}_{H+} + \lambda^{\infty}{}_{Cl-} = 426\times10^{-4}\ \text{S m}^2\ \text{mol}^{-1}$$

$$\alpha = \frac{\Lambda_m - \Lambda'_m}{\Lambda_{HCl} - \Lambda'_m} = \frac{(1.50-1)\times10^{-2}\ \text{S m}^2\ \text{mol}^{-1}}{(4.26-1)\times10^{-2}\ \text{S m}^2\ \text{mol}^{-1})} = 0.1534$$

$$K_h = \alpha^2 c/(1-\alpha) = (0.1534)^2\times0.001/(1-0.1534) = 2.78\times10^{-5}$$

9.9 Experimental Determination of Solubility Products

A sparingly soluble salt in solution forms a saturated solution in which the solid is in equilibrium with the dissolved ions. For example, consider the salt silver chloride. The solubility product of silver chloride is given by

$$K_{sp} = [Ag^+][Cl^-] \tag{9.48}$$

It can be assumed that the ions are formed in such low concentrations that their activity coefficients can be taken as unity. If *S* is the solubility of AgCl,

$$K_{sp} = (S/C^0)^2 \tag{9.49}$$

where C^0 is standard concentration (1 mol dm^{-3}), a term introduced to make K_{sp} dimensionless.

The conductivity of the solution is due to the dissolved ions and the dissociated water molecules.

$$\chi = C_{Ag+}\lambda_{Ag+} + C_{Cl-}\lambda_{Cl-} + C_{H+}\lambda_{H+} + C_{OH-}\lambda_{OH-} \quad (9.50)$$

Since ionisation of water is not affected by the presence of sparingly soluble salt, the last two terms represent the conductivity of water, K_w.

$$\chi - K_w = C_{Ag+}\lambda_{Ag+} + C_{Cl-}\lambda_{Cl-} \quad (9.51)$$

If the concentration of ions formed by dissolution of silver chloride is less than 10^{-3} N, the molar conductivities of the ions can be assumed to be equal to their infinite dilution values.

$$\chi - K_w = C_{Ag+}\lambda^{\infty}_{Ag+} + C_{Cl-}\lambda^{\infty}_{Cl-} \quad (9.52)$$

Since AgCl is a 1:1 electrolyte, its solubility S will be equal to the concentrations of the ions formed.

$$C_{Ag+} = C_{Cl-} = S$$

$$\chi - K_w = S(\lambda^{\infty}_{Ag+} + \lambda^{\infty}_{Cl-}) \quad (9.53)$$

$$S = \frac{\chi - K_w}{(\lambda^{\infty}_{Ag+} + \lambda^{\infty}_{Cl-})}$$

From equation (9.49),

$$S = C^0 K_{sp}^{1/2} \quad (9.54)$$

$$K_{sp} = \frac{(\chi - K_w)^2}{(C^0)^2(\lambda^{\infty}_{Ag+} + \lambda^{\infty}_{Cl-})^2} \quad (9.55)$$

As an example, if $\chi - K_w = 1.802 \times 10^{-4}$ S m^2 mol^{-1} and $(\lambda^{\infty}_{Ag+} + \lambda^{\infty}_{Cl-})$ $= 138.27 \times 10^{-4}$ S m^2 mol^{-1},

$$K_{sp} = \frac{(1.802 \times 10^{-4}\ \text{S m}^2\ \text{mol}^{-1})^2}{(1\ \text{mol dm}^{-3} \times 10^3\ \text{m}^{-3}/\text{dm}^{-3})^2\ (138.27 \times 10^{-4}\ \text{S m}^2\ \text{mol}^{-1})^2}$$

$$= 1.69 \times 10^{-10}.$$

9.10 Conductometric Titrations

The end point in titrations of acids and bases, or precipitation reactions can be determined by conductivity measurements.

9.10.1 Titration of strong acid with strong base

Consider the titration of NaOH with HCl. The conductivity cell is placed in the NaOH solution and HCl is added from the burette. Initially the conductance of the solution is entirely due to Na^+ and OH^- ions from NaOH. When a drop of HCl is added from the burette, some of the OH^- ions are removed due to neutralisation.

$$(Na^+ + OH^-) + (H^+ + Cl^-) \rightarrow H_2O + Na^+ + Cl^-$$

This equation shows that, in effect, OH^- ions have been replaced by Cl^- ions. Since the ionic molar conductivity of OH^- ions (198×10^{-4} S m^2 mol^{-1})

is much greater than that of Cl^- ions (78.3×10^{-4} S m^2 mol^{-1}), there is a net decrease in the conductivity of the solution. Thus with the addition of each drop of HCl to the solution, there is a progressive decrease in the conductivity of the solution. The decrease continues till the end point of the titration. At the end point, all the OH ions have been neutralised by the H^+ ions. With the next drop of acid added, the H^+ and Cl^- ions from the acid remain in solution and contribute to a sharp rise in conductivity.

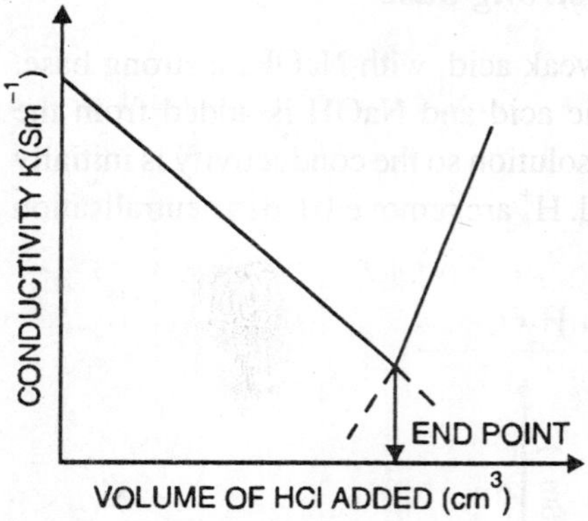

Figure 9.5 (a): Titration of NaOH with HCl

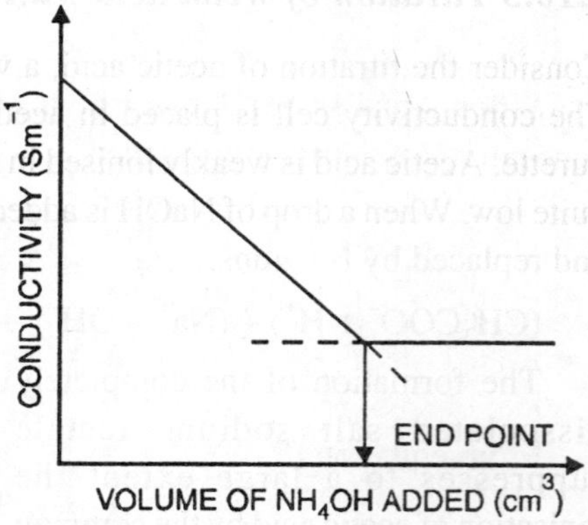

Figure 9.5(b): Titration of NH_4OH with HCl

The variation in the conductivity of the solution with volume of the acid added is shown in Figure 9.5(a). Knowing the normality of the acid taken and the volume required to neutralise the known volume of NaOH taken in the conductivity cell, the normality of the NaOH solution can be calculated.

9.10.2 Titration of weak base with strong acid

Consider the titration of a weak base like ammonium hydroxide with a strong acid like hydrochloric acid. The conductivity cell is placed in HCl while NH_4OH is added from the burette. Initially the conductivity of the solution is high due to the presence of H^+ and Cl^- ions from HCl. As NH_4OH is added to the solution, H^+ ions are neutralised and replaced by NH_4^+ ions.

$$(H^+ + Cl^-) + (NH_4^+ + OH^-) \rightarrow H_2O + NH_4^+ + Cl^-$$

Since the ionic molar conductivity of NH_4^+ ions (73.5×10^{-4} S m^2 mol^{-1}) is much less than that of H^+ ions (349×10^{-4} S m^2 mol^{-1}), there is a net decrease in the conductivity of the solution. This decrease continues till the end point of the titration. After the end point, addition of excess of NH_4OH does not cause the conductivity to change appreciably since the base is very weakly dissociated.

The variation in the conductivity of the solution with volume of the weak base added is shown in Figure 9.5(b). Near the end point, a slight curvature is seen. It occurs because ammonium chloride formed as a product of the reaction hydrolyses to form H_3O^+ ions.

$$NH_4^+Cl^- + H_2O \longleftrightarrow NH_4OH + H_3O^+ + Cl^-$$

The ionic molar conductivity of H_3O^+ being much higher than that of NH_4^+ ions, a slight increase in the conductance is observed.

Knowing the normality of NH_4OH taken and the volume required to neutralise the known volume of HCl taken in the conductivity cell, the normality of the HCl solution can be calculated.

9.10.3 Titration of weak acid with strong base

Consider the titration of acetic acid, a weak acid, with NaOH, a strong base. The conductivity cell is placed in acetic acid and NaOH is added from the burette. Acetic acid is weakly ionised in solution so the conductivity is initially quite low. When a drop of NaOH is added, H^+ are removed due to neutralisation and replaced by N ions.

$$(CH_3COO^- + H^+) + (Na^+ + OH^-) \rightarrow H_2O + CH_3COO^- + Na^+$$

The formation of the completely dissociated salt sodium acetate suppresses to a large extent the ionisation of acetic acid by the common ion effect. This leads to a net decrease in the conductance of the solution initially. As more NaOH is added the number of CH_3COO^- and Na^+ ions increases, which raises the conductivity of the solution. A slow rise in conductivity continues till the end point is reached. After the end point, the addition of excess NaOH leads to a sharp rise in conductivity.

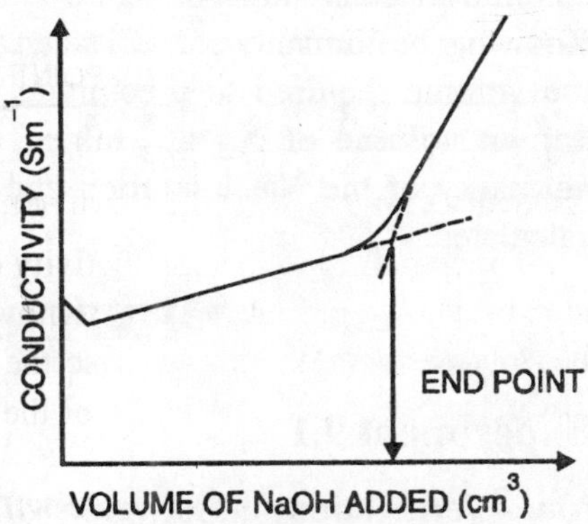

Figure 9.5(c): Titration of acetic acid with NaOH.

The variation in the conductivity of the solution with volume of the strong base added is shown in Figure 9.5 (c). A slight curvature is seen near the end point. This is due to the hydrolysis of sodium acetate.

$$CH_3COO^- + Na^+ + H_2O \rightarrow CH_3COOH + H_3O^+ + Na^+$$

The fast moving H_3O^+ ions raise the conductivity, leading to conductivity values slightly higher than expected near the end point.

Knowing the normality of NaOH taken and the volume required to neutralise the known volume of acetic acid taken, the normality of the acetic acid solution can be calculated.

9.10.4 Precipitation titrations

Consider the titration of silver nitrate with sodium chloride. Silver nitrate solution is taken in the conductivity cell and sodium chloride is added from the burette. When a drop of sodium chloride is added silver chloride is formed, which precipitates out from the solution.

$$(Ag^+ + NO_3^-) + (Na^+ + Cl^-) \rightarrow AgCl\downarrow + Na^+ + NO_3^-$$

The net effect of the reaction is the replacement of Ag^+ ions with Na^+ ions. The ionic molar conductivity of Ag^+ (61.9×10^{-4} S m^2 mol^{-1}) being only slightly more than that of Na^+ (50.1×10^{-4} S m^2 mol^{-1}), there is a slight net decrease in the conductivity of the solution. The decrease in conductivity continues with further addition of NaCl till the end point. After the end point, the presence of excess Na^+ and Cl^- ions from NaCl causes a sharp rise in the conductivity of the solution.

The variation in the conductivity of the solution with volume of the sodium chloride added is shown in Figure 9.5(d). Knowing the normality of NaCl taken and the volume required to precipitate the known volume of $AgNO_3$ taken, the normality of the NaCl solution can be calculated.

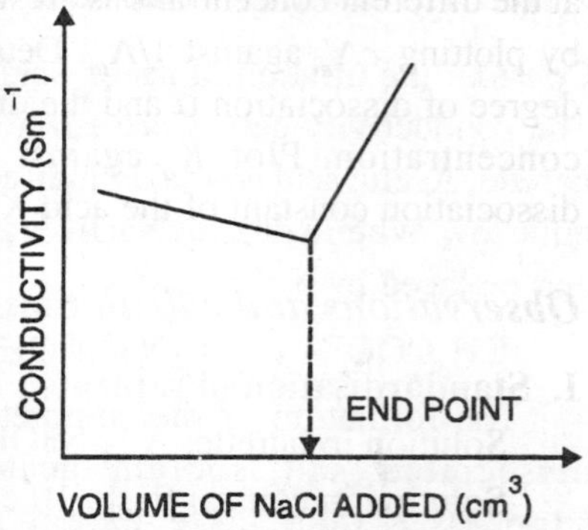

Figure 9.5 (d): Titration of silver nitrate with sodium chloride.

EXPERIMENTS

Experiment 9.1

Aim: Verification of Ostwald's dilution law and determination of the dissociation constant of a weak monobasic acid conductometrically.

Requirements

Direct reading conductivity meter, conductivity cell, conductivity water, 0.01M KCl, benzoic acid (AR grade), 0.05M NaOH, 0.05M succinic acid solution.

Procedure

Prepare a saturated solution of benzoic acid at about 40°C. Allow it to cool at room temperature and filter through Whatmann filter paper number 1 into a dry bottle. Determine the concentration of this solution by titrating 25 cm^3 of this solution with previously standardised 0.05M NaOH. Dilute the solution appropriately so as to get 250 cm^3 of 0.02M benzoic acid solution. Prepare five more solutions of concentration 0.01M, 0.005M, 0.0025M, 0.00125M by appropriately diluting the 0.02M benzoic acid solution.

Determine the cell constant of the conductance cell using 0.1M KCl solution. (With direct reading digital conductivity meters, calibrate the instrument to get full-scale reading and then standardise for the given cell using 0.1M KCl solution.) The variation of conductivity of 0.1M KCl with temperature can be looked up in Table 9.4 in Appendix 9.

Measure the conductivity of each benzoic acid solution, preferably starting with the most dilute solution. Finally determine the conductivity of the water used throughout the experiment. Subtract the conductivity of water from the conductivity of each solution and evaluate the molar conductivity of the acid at the different concentrations. Test the validity of the Ostwald's Dilution Law by plotting $c\Lambda_m$ against $1/\Lambda_m$. Determine $\Lambda_m°$ from the graph. Calculate the degree of dissociation α and the dissociation constant K_a' of the acid at each concentration. Plot K_a against $\sqrt{(\alpha c)}$. Determine the thermodynamic dissociation constant of the acid K_a.

Observations and calculations

I. Standardisation of saturated benzoic acid solution

Solution in burette: **A** N NaOH (i.e., previously standardised)

Solution in conical flask: 10 cm^3 saturated benzoic acid solution

Indicator used: phenolphthalein

End point: colourless to pale pink

	Burette readings (cm^3)			Constant burette reading
	(I)	(II)	(III)	
Final Initial Difference				**V** cm^3

$$\begin{aligned} \text{NaOH} &= C_6H_5COOH \\ N_1V_1 &= N_2V_2 \\ \mathbf{A\ V} &= N_2 \times 10 \\ N_2 &= \mathbf{AV}/10 \end{aligned}$$

Normality of saturated benzoic acid solution = ______ N

II. Determination of conductivity of benzoic acid solutions

Conductivity of distilled water = ______ S

Conc. of benzoic acid c (M)	Conductivity χ (S cm^{-1})	Corrected conductivity χ (S cm^{-1})	Molar conductivity λ = 1000 χ/c (S cm^2 mol^{-1})	$1/\lambda$
0.00125 ... 0.02				

$c\lambda$	Degree of dissociation $\alpha = \lambda / \lambda_\infty$	Dissociation constant $K_a' = a^2 c/(1-\alpha)$	$\log K_a'$	$\sqrt{c}$

Mean K_a' = ______

Therefore, dissociation constant (by calculations) = mean K_a' = ______

From the graph of $c\lambda$ versus $1/\lambda$,

Slope = ______ and y-intercept = ______

Equivalent conductance at infinite dilution λ_∞ = slope/intercept = ______

From the graph of $\log K_a'$ versus $\sqrt{c}$,

y-intercept = ______ = $\log K_a$

K_a = antilog (y-intercept) = ______

Therefore, thermodynamic dissociation constant = K_a = ______

Result

The dissociation constant K_a of benzoic acid is

(i) by graph (thermodynamic dissociation constant) = ______

(ii) by calculation = ______

Experiment 9.2

Aim: To determine the hydrolysis constant of a salt of strong acid and weak base e.g. aniline hydrochloride.

Requirements

Aniline hydrochloride A.R. (recrystallised), aniline A.R. (redistilled), conductometer, conductivity cell.

Procedure

Prepare 1/32 mol dm^{-3} solution of aniline hydrochloride in water. Talking all the usual precautions of rinsing the cell, measure the conductivity of the solution. Dilute the solution appropriately to give 1/64 mol dm^{-3} and 1/128 mol dm^{-3} solution in water and determine the conductivity of each solution.

Now prepare 1/32 mol dm^{-3} solution of aniline hydrochloride in 1/32 dm^{-3} solution of aniline and determine the conductivity of this solution. Dilute this solution with aniline to give 1/64 and 1/128 mol dm^{-3} solutions and measure the conductivity of each dilution.

Observations and Calculations

Aniline hydrochloride in water

Concentration c / (mol dm^{-3})	Conductivity χ / (S cm^{-1})	Molar conductivity $\Lambda_m = 1000\ \chi/c$ /(S cm^2 mol^{-1})
1/32 1/64 1/128		

Aniline hydrochloride in aniline

Concentration c / (mol dm^{-3})	Conductivity χ / (S cm^{-1})	Molar conductivity $\Lambda_m = 1000\ \chi/c$ /(S cm^2 mol^{-1})
1/32 1/64 1/128		

Hydrochloric acid

Concentration c / (mol dm^{-3})	Conductivity χ / (S cm^{-1})	Molar conductivity $\Lambda_m^\circ = 1000\ \chi/c$ /(S cm^2 mol^{-1})
1/32 1/64 1/128		

Concentration c / (mol dm^{-3})	$\Lambda^\circ_m - \Lambda'_m$ /(S cm^2 mol^{-1})	$\Lambda_m - \Lambda'_m$ /(S cm^2 mol^{-1})	$h = (\Lambda_m - \Lambda'_m)/(\Lambda^\circ_m - \Lambda'_m)$	$K_h = h^2c/(1-h)$ / (mol dm^{-3})

Mean K_h = —— mol dm^{-3}

Dissociation constant of aniline $K_b = K_w/K_h$, where K_w is the ionic product of water.

Result

1. Hydrolysis constant of aniline hydrochloride = _____ mol dm^{-3}
2. Dissociation constant of weak base aniline = _____ mol dm^{-3}

Experiment 9.3

Aim: Titration of a mixture of trichloroacetic acid, monochloroacetic acid and acetic acid by titrating with sodium hydroxide conductometrically.

Theory

When a mixture of trichloroacetic acid, monochloroacetic acid and acetic acid is titrated with sodium hydroxide, the strongest acid, namely trichloroacetic acid, is neutralised first and is accompanied by the decrease in conductance of the solution. This is followed by neutralisation of monochloroacetic acid and it is seen that there is not so sharp a decrease in conductance than the previous one. Upon further addition of the base, the conductance of the solution increases as acetic acid is replaced by its salt and the final step rise in conductance is due to the free ions of OH^-. The extrapolation of the four straight portions of the graph of conductance versus volume of sodium hydroxide added leads to the location of end point of the titration.

Requirements

Conductivity cell, 0.25 N NaOH, mixture of trichloroacetic acid, monochloroacetic acid and acetic acid.

Procedure

Pipette out 50 cm^3 of the mixture of acids into a 100 cm^3 beaker. Immerse the cell in this solution and determine the conductance of the solution. Add 0.5 cm^3 of previously standardised 0.25 N NaOH solution. Stir well and determine the conductance. Repeat, stirring well after each addition.

With direct reading conductivity meter, calibrate the instrument to get full-scale reading and then standardise for the given cell using 0.1 M potassium chloride solution.

To apply correction for the dilution effect, multiply each conductance/ conductivity value by $(V + v)/V$ where V is the original volume of solution and v the volume of sodium hydroxide added. Plot corrected conductance/ conductivity values against the volume of sodium hydroxide solution added and extrapolate the straight-line portions of the graph to give the end points. Calculate the concentration and amount/weight of each acid in the given solution.

Observations and calculations

Solution in burette: 0.25 N NaOH

Solution in conductivity cell: 50 cm^3 of the mixture of acids

Vol of NaOH added (cm^3)	Conductivity χ(S cm^{-1})	Volume corrected conductivity $\chi\ (V + v)/V$ (S cm^{-1})
0.0		
0.5		
1.0		
...		

From the graph of Volume corrected conductivity versus volume of NaOH added,

$V_1 =$ ______ cm^3, $V_2 =$ ______ cm^3, $V_3 =$ ______ cm^3

Volume of NaOH required to titrate trichloroacetic acid = V_1 cm^3

Volume of NaOH required to titrate monochloroacetic acid = $V_2 - V_1$ cm^3

Volume of NaOH required to titrate acetic acid = $V_3 - V_2$ cm^3

Result

Acid	Concentration (mol dm^{-3})	Amount in the given soln (mol)	Weight in the given soln (g)
Trichloroacetic acid			
Monochloroacetic acid			
Acetic acid			

Experiment 9.4

Aim: Determination of the concentrations/amounts of sulphuric acid, acetic acid and copper sulphate by conductometric titration with sodium hydroxide.

Theory

When a mixture of strong acid, weak acid and a salt is titrated against strong alkali, the strong acid is neutralised first. This results in a rapid fall in conductance as fast moving H^+ ions are replaced by slower moving hydroxide ions. After the strong acid has been neutralised, the weak acid is neutralised next. The conductance rises slowly due to the hydrolysis of the salt formed. Finally, sodium hydroxide reacts with the salt. In case of copper sulphate, a precipitation reaction takes place as follows:

$$CuSO_4 + 2NaOH \rightarrow Cu(OH)_2\downarrow + Na_2SO_4$$

The Cu^{2+} ions are replaced by slightly less mobile Na^+ ions and hence conductance decreases very slowly until the precipitation is complete. After the end point, the conductance rises very rapidly due to the addition of fast moving OH^- ions. A plot of conductance of the solution versus volume of the titrant added will show three breaks. The three intersection points give the volume of NaOH required to titrate the strong acid, the weak acid and the salt respectively.

Requirements

Direct reading conductivity meter, conductivity cell, 0.1N H_2SO_4, 0.1N CH_3COOH, 0.1N $CuSO_4.5H_2O$ and 0.25N NaOH.

Procedure

Determine the cell constant of the conductance cell using 0.1M KCl solution. (With direct reading digital conductivity meters, calibrate the instrument to get full-scale reading and then standardise for the given cell using 0.1M KCl solution.) The variation of conductivity of 0.1M KCl with temperature can be looked up in Table 9.4 in Appendix 9.

Pipette out 10 cm^3 of H_2SO_4 solution, 10 cm^3 of CH_3COOH solution, and 10 cm^3 of $CuSO_4.5H_2O$ solution in a 250 cm^3 beaker. Dip the conductance cell in the beaker. Measure the conductivity of the solution. Fill a burette with previously standardised 0.25N NaOH solution. Titrate the mixture conductometrically. Note down the conductance value of the solution after each addition of 0.5 cm^3 of standard NaOH solution from the burette.

To apply correction for the dilution effect, multiply each conductance/conductivity value by $(V + v)/V$ where V is the original volume of solution and v the volume of sodium hydroxide added. Plot corrected conductance/conductivity values against the volume of sodium hydroxide solution added. Extrapolate the straight-line portions of the graph to give the end points. Calculate the concentration and amount/weight of sulphuric acid, acetic acid and copper sulphate in the given solution.

Observations and calculations

Solution in burette: 0.25 N NaOH

Solution in conductivity cell: 30 cm^3 of mixture of sulphuric acid, acetic acid and copper sulphate solution

Vol. of NaOH added (cm^3)	Conductivity k ($S\ cm^{-1}$)	Volume corrected conductivity $k(V + v)/V$ ($S\ cm^{-1}$)
0.0		
0.5		
1.0		
...		

From the graph of volume corrected conductivity versus volume of NaOH added,

V_1 = _____ cm^3, V_2 = _____ cm^3, V_3 = ______ cm^3

Volume of NaOH required to titrate sulphuric acid = V_1 cm^3

Volume of NaOH required to titrate acetic acid = $V_2 - V_1$ cm^3

Volume of NaOH required to titrate copper sulphate = $V_3 - V_2$ cm^3

Result

Substance	Concentration (mol dm^{-3})	Amount in the given soln (mol)	Weight in the given soln (g)
Sulphuric acid			
Acetic acid			
Copper sulphate			

Experiment 9.5

Aim: Determination of the concentration/amount of oxalic acid by conductometric titration with sodium hydroxide.

Theory

Oxalic acid is a dibasic acid. It dissociates in two steps and thus has two dissociation constants. The first dissociation is similar to that of a strong acid and the second one similar to that of a weak acid. Since the two dissociation constants are well separated, they can be easily detected by conductometric titration. At the start of the titration the conductance decreases rapidly till the first dissociation is complete and then rises slowly up to the second dissociation. After the second dissociation is complete, the conductance rises rapidly due to the addition of fast moving OH^- ions added due to excess NaOH. The titration curve will contain two straight-line portions.

Requirements

Direct reading conductivity meter, conductivity cell, 0.1N oxalic acid, 0.25N NaOH solution.

Procedure

Determine the cell constant of the conductance cell using 0.1M KCl solution. (With direct reading digital conductivity meters, calibrate the instrument to get full-scale reading and then standardise for the given cell using 0.1M KCl solution.) The variation of conductivity of 0.1M KCl with temperature can be looked up in Table 9.4 in Appendix 9.

Pipette out 20 cm^3 of oxalic acid solution in a 150 cm^3 beaker. Add 30 cm^3 of distilled water and dip the conductivity cell in the solution. Measure the conductivity of the solution. Fill a burette with previously standardised 0.25N NaOH solution. Titrate the mixture conductometrically. Note down the conductance value of the solution after each addition of 0.5 cm^3 of standard NaOH solution from the burette.

To apply correction for the dilution effect, multiply each conductance/conductivity value by $(V + v)/V$ where V is the original volume of solution and

V the volume of sodium hydroxide added. Plot corrected conductance/conductivity values against the volume of sodium hydroxide solution added. Extrapolate the straight-line portions of the graph to give the end points. Calculate the concentration and amount/weight of oxalic acid in the given solution.

Observations and calculations

Solution in burette: 0.25 N NaOH

Solution in conductivity cell: 20 cm^3 of oxalic acid solution

Vol. of NaOH added (cm^3)	Conductivity χ (S cm^{-1})	Volume corrected conductivity $\chi(V + v)/V$ (S cm^{-1})
0.0		
0.5		
1.0		
...		

From the graph of volume corrected conductivity versus volume of NaOH added,

V_1 = _____ cm^3, V_2 = _____ cm^3

Result

1. Concentration of oxalic acid in the given solution = _____N
2. Amount of oxalic acid in the given soln = _____mol
3. Weight of oxalic acid in the given soln = _____g

Experiment 9.6

Aim: To determine the transport number of copper ions in copper sulphate by Hittorf's method.

Requirements

Transport number cell with copper electrodes, silver coulometer, DC source, connecting wires, $CuSO_4$ solution, standard 0.1M $Na_2S_2O_3$ soln, solid KI, starch solution, 0.1M $K_2Cr_2O_7$.

Procedure

Prepare 500 cm^3 of 0.05M $CuSO_4$ soln in distilled water. Clean the transport number cell and dry it; rinse with the copper sulphate solution. Insert the copper electrodes in the two limbs.

Clean the platinum crucible of the silver coulometer by dilute nitric acid and then wash with distilled water. Dry and weigh accurately. Fill the coulometer with 15% silver nitrate and insert a silver electrode.

Connect the transport number cell in series with the silver coulometer. Adjust the current strength to about 30 mA. Allow the current to pass for two hours so that 100 coulombs of electricity is passed.

In the meantime prepare 250 cm^3 of 0.1M $Na_2S_2O_3$ and standardise it with 0.1M $K_2Cr_2O_7$ solution. Take 25 cm^3 of $CuSO_4$ solution in a dry, weighed conical flask. Reweigh it and titrate with standardised 0.1M $Na_2S_2O_3$ solution and determine the amount of copper sulphate.

At the end of the electrolysis, close the current flow. With draw the whole anodic compartment solution in a weighed flask and reweigh it and titrate it with standard 0.1M $Na_2S_2O_3$ solution using starch as the indicator.

Remove the platinum crucible, wash it gently with distilled water, dry and weigh it. The drying can be done by soaking up droplets of solution with Whatmann filter paper number 1, and then heating the crucible in the oven at 110°C.

From the titration, calculate the amount of copper and water in the anolyte compartment. Calculate the amount of copper in this quantity of solution before electrolysis. Calculate the amount of copper equivalent to silver in the coulometer. Hence determine the transport number of Cu^{2+} ions as

$$t_{Cu2+} = \frac{\text{loss in amount of copper in anolyte due to migration}}{\text{copper equivalent to silver in coulometer}}$$

Observations and calculations

[A] Standardisation of 0.1M (approx) $Na_2S_2O_3$

Solution in burette: 0.1M (approx) $Na_2S_2O_3$ solution

Solution in conical flask: 10 cm^3 of 0.1N $K_2Cr_2O_7$ + 5 cm^3 conc. HCl + solid KI

Indicator: starch solution

End point: dark blue to colourless

	Burette readings (cm^3)			Constant burette reading
	(I)	(II)	(III)	
Final Initial Difference				V cm^3

$$Na_2S_2O_3 = K_2Cr_2O_7$$
$$N_1V_1 = N_2V_2$$
$$N_1 = \_\_\_\_ N$$

[B] Before electrolysis

Wt of empty conical flask = W_1 = ____g

Wt of conical flask + 10 cm^3 0.05M $CuSO_4$ = W_2 = ____g

Wt of $CuSO_4$ soln = $W_2 - W_1$ = ____g

Volume of N_1 M $Na_2S_2O_3$ required to titrate 10 cm^3 0.05M $CuSO_4$ = A cm^3 = ____ cm^3

Now 1000 cm^3 of 1M $Na_2S_2O_3$ ≡ 159.64 g of $CuSO_4$

Therefore, A cm^3 of N_1 M $Na_2S_2O_3$ ≡ 159.64 N_1 A /1000 = W_3 = ____ g of $CuSO_4$

Now W_3 g of $CuSO_4$ is present in $W_2 - W_1$ g of solution.

Therefore, amount of water present = $W_3 - (W_2 - W_1)$ = ____ g

[C] After electrolysis

Wt of empty conical flask = W_4 = ____ g

Wt of conical flask + anolyte solution = W_5 = ____ g

Wt of anolyte solution = $W_5 - W_4$ = ____ g

Volume of N_1 M $Na_2S_2O_3$ required to titrate anolyte solution = B cm^3 = ____ cm^3

Now 1000 cm^3 of 1 M $Na_2S_2O_3$ = 159.64 g of $CuSO_4$

Therefore, B cm^3 of N_1 M $Na_2S_2O_3$ = 159.64 N_1 B /1000 = W_6 = ____ g of $CuSO_4$

Now W_6 g of $CuSO_4$ is present in $W_5 - W_4$ g of solution.

Therefore, amount of water present = $W_6 - (W_5 - W_4)$ = ____ g

Before electrolysis, $W_3 - (W_2 - W_1)$ g of water contained W_3 g of $CuSO_4$

Therefore, $W_6 - (W_5 - W_4)$ g of water should contain

$[W_6 - (W_5 - W_4)]\ W_3 / [W_3 - (W_2 - W_1)] = W_7$ g of $CuSO_4$

But $W_6 - (W_5 - W_4)$g of water contains W_6 g of $CuSO_4$

Therefore, increase in amount of $CuSO_4$ = $W_6 - W_7$ g

Now 159.64 g of $CuSO_4$ = 63.54 g of Cu

Therefore $W_6 - W_7$ g of $CuSO_4$ = 63.54 $(W_6 - W_7)$/159.64 g of $CuSO_4$ = W_8 g of $CuSO_4$

[D] From coulometer

Wt of empty Pt crucible = W_9 = ____ g

Wt of Pt crucible + silver deposited = W_{10} = ____ g

Wt of silver deposited = $W_{10} - W_9$ = ____ g

Now 107.88 g of silver = 31.78 g of Cu deposited

Therefore, $W_{10} - W_9$ g of silver = 31.78 $(W_{10} - W_9)$/ 107.88 g of Cu = W_{11} = ____ g

[E] Transport number

$$t_{Cu2+} = (W_{11} - W_8)/W_{11} = \_\_\_\_$$

$$t_{SO4\ 2-} = 1 - t_{Cu2+} = \_\_\_\_$$

Result

Transport number of Cu^{2+} ions in $CuSO_4$ soln t_{Cu2+} = ____

Transport number of SO_4^{2-} ions in $CuSO_4$ soln $t_{SO4\ 2-}$ = ____

Experiment 9.7

Aim: To determine the transport number of ions in solution using moving boundary method.

Procedure

Set-up the apparatus for the moving boundary method, as shown in Figure 9.3.

Fill the capillary with standardised 0.1 N HCl to which a small quantity of methyl orange indicator has been added. Connect it to a constant current source. In case one is not available, use a coulometer to measure the quantity of electricity passed through the cell. Pass current though the cell. Cadmium dissolves at the anode, forming cadmium chloride. Cd^{2+} and H^+ ions migrate in the upward migration. A boundary can be clearly distinguished as these ions move towards the cathode. The solution above the boundary is pinkish in colour because it contains the fast-moving H^+ ions. The solution below the boundary is yellowish-orange in colour.

Taking the time of starting the current as zero time, find the time taken to sweep a total volume of 0.20 cm^3, taking readings every 0.02 cm^3. Plot a graph of volume swept versus time taken and find its slope. The transport number of hydrogen ion can be calculated from the formula

$$t_{H+} = \frac{V}{t}\frac{cF}{I} \tag{9.56}$$

where V is the volume in cm^3 swept in t seconds

V/t is the slope of the graph of volume versus time

c is the concentration of HCl used

F is the Faraday constant

I is the current passed in amperes

Observations and calculations

Standardisation of HCl

Solution in burette: 0.1 N standard NaOH solution

Solution in conical flask: 10 cm^3 0.1 N (approx) HCl

Indicator: phenolphthalein

End point: colourless to pink

	Burette readings (cm^3)			*Constant burette reading*
	(I)	(II)	(III)	
Final				
Initial				— cm^3
Difference				

NaOH ≡ HCl

$$N_1V_1 = N_2V_2$$

$$N_2 = \_\_\_\_\_ \text{ N}$$

Thus the concentration of HCl is N_2.

Determination of transport number

Current passed = ______ A

Volume swept	Time elapsed	
V/ (cm^3)	In minutes t / (min)	In seconds t / (s)
0.00	0	0
...		
0.20		

From the graph of volume swept versus time taken, slope = _____

Substituting in equation (9.56),

$$t_{H+} = \frac{\text{slope} \times N_2 F}{I}$$

Result

The transport number of hydrogen ion as determined by the moving boundary method is ____ .

Appendix

Review Questions

1. How will you construct a conductance cell with unit cell constant?
2. The transport number of cadmium ions in cadmium iodide solution varies markedly with concentration. At 0.01M concentration, it is 0.444, at 0.1 M it is 0.296, and at 0.5 M it becomes 0.003. At concentrations greater than 0.5 M, the transport number of cadmium ions apparently becomes negative. Explain.
3. What is the unit of thermodynamic solubility product?
4. What is the necessity of correcting conductance values during a conductometric titration?
5. Why should conductivity water be free of carbon dioxide?

Table 9.1: Ionic molar conductivity at infinite dilution in water at 298 K

Cation	λ^∞ (10^{-4} S m^2 mol^{-1})	Anion	λ^∞ (10^{-4} S m^2 mol^{-1})
H^+	349.81	OH^-	198.3
Li^+	38.68	F^-	55.4
Na^+	50.10	Cl^-	76.35
K^+	73.50	Br^-	78.14

Rb^+	77.81	I^-	76.84
Cs^+	77.26	NO_3^-	71.46
Ag^+	61.90	ClO_3^-	64.6
NH_4^+	73.55	BrO_3^-	55.74
$[(CH_3)_4]N^+$	44.92	IO_3^-	40.54
$[(C_2H_5)_4]N^+$	32.66	ClO_4^-	67.36
$[(C_3H_7)_4]N^+$	23.42	IO_4^-	54.55
Be^{2+}	90	HCO_3^-	44.50
Mg^{2+}	106.10	$HCOO^-$	54.59
Ca^{2+}	119.00	CH_3COO^-	40.90
Sr^{2+}	118.90	CH_2BrCOO^-	39.22
Ba^{2+}	127.26	$(NO_2)_3C_6H_2O^-$	30.39
Cu^{2+}	107.2	SO_4^{2-}	160.04
Zn^{2+}	105.6	$C_2O_4^{2-}$	148.30
Co^{2+}	110	CO_3^{2-}	138.6
Pb^{2+}	139.0	$[Fe(CN)_6]^{3-}$	302.7
La^{3+}	209.1	$P_3O_9^{3-}$	250.8
Ce^{3+}	209.4	$[Fe(CN)_6]^{4-}$	442.0
$[Co(NH_3)_6]^{3+}$	305.7	$P_4O_{12}^{4-}$	374.8
$[Ni\textit{tri–en}_3]^{4+}$	210.0	$P_2O_7^{4-}$	383.6
$[Co_2\textit{tri–en}_3]^{6+}$	412.2	$P_3O_{10}^{5-}$	545

Table 9.2: Ionic mobility in water at 298 K

Cation	*u* (10^{-8} m^2 s^{-1} V^{-1})	Anion	*u* (10^{-8} m^2 s^{-1} V^{-1})
H^+	36.23	OH^-	20.64
Li^+	4.01	F^-	5.70
Na^+	5.19	Cl^-	7.91
K^+	7.62	Br^-	8.09
Rb^+	7.92	I^-	7.96
Ag^+	6.42	NO_3^-	7.40
NH_4^+	7.63	CH_3COO^-	4.24
$[(CH_3)_4]N^+$	4.65	SO_4^{2-}	8.29
Ca^{2+}	6.17	CO_3^{2-}	7.46
Cu^{2+}	5.56	$[Fe(CN)_6]^{3-}$	10.5
Zn^{2+}	5.47	$[Fe(CN)_6]^{4-}$	11.4

Table 9.3: Conductivity of KCl in water

Temperature of solution/ (°C)	Conductivity / (S m^{-1})	
	0.1 M KCl	0.01 M KCl
15	0.01048	0.001147
16	0.01072	0.001173
17	0.01095	0.001199
18	0.01119	0.001225
19	0.01143	0.001251
20	0.01167	0.001278
21	0.01191	0.001305
22	0.01215	0.001332
23	0.01239	0.001359
24	0.01264	0.001386
25	0.01288	0.001413
26	0.01313	
27	0.01337	
28	0.01362	
29	0.01387	
30	0.01412	

Table 9.4: Transport numbers for aqueous solutions at 298 K

Electrolyte	Concentration (moles of positive or negative charge per litre)			
	0.01	0.05	0.1	0.2
HCl	0.8251	0.8292	0.8314	0.8337
NaCl	0.3918	0.3876	0.3854	0.3821
KCl	0.4902	0.4899	0.4898	0.4894
NH_4Cl	0.4907	0.4905	0.4907	0.4911
KNO_3	0.5084	0.5093	0.5103	0.5120
Na_2SO_4	0.3848	0.3829	0.3828	0.3828
K_2SO_4	0.4829	0.4870	0.4890	0.4910

Expected Experimental Results

Experiment 9.1

Dissociation constant of benzoic acid is 6.46×10^{-5}.

Experiment 9.2

Hydrolysis constant of aniline hydrochloride is 2.5×10^{-5} mol dm^{-3}.

Experiment 9.6

Transport number of copper ions in copper sulphate is 0.38. Transport number of silver ions in silver nitrate is 0.47. Transport number of hydrogen ions in nitric acid is 0.84.

Experiment 9.7

Transport number of hydrogen ions is 0.84.

Hints to Review Questions

1. A conductivity cell, whose electrodes have a surface area of 1 cm^2 each and are 1 cm distance apart, will have unit cell constant.
2. A negative transport number indicates that cadmium is being transported in a direction opposite to that in which positive charge moves through the solution. In dilute solutions, cadmium iodide dissociates to give Cd^{2+} ions and I^- ions but in concentrated solutions, the I^- ions combine with unionised molecules of CdI_2 to form the negatively charged complex $[CdI_4]^{2-}$. Since cadmium is part of the negative ion, its transport number is apparently negative. Such a transport number is called an *abnormal transport number*.
3. The thermodynamic solubility product is a mathematical product of the activities of the ions in solution. Since activity itself is a dimensionless quantity, the thermodynamic solubility product also has no units.
4. The conductance of a solution varies with its concentration. As a conductometric titration proceeds the titre solution gets diluted due to the volume of reagent added. Hence correction for dilution is required. This correction can be avoided if the titrant is much more concentrated than the titre solution.
5. Carbon dioxide dissolves in water to form carbonic acid. The ions produced due to the weak dissociation of carbonic acid boost the conductivity readings, thus giving rise to errors.

Further Exploration through the Internet

http://www–ec.njit.edu/~grow/conductivity.htm

http://physchemlab.tau.ac.il/experiments/Conductivity/conductivity.htm

http://chemistry.semo.edu/pchem/lab/expt21/expt21.html

http://tardis.csudh.edu/fromero/chemistry/condu/

Luigi Galvani

Luigi Galvani was born on September 9, 1737 in Bologna, Italy. He studied medicine at the University of Bologna. In 1762 he became Reader there in anatomy and surgery. In a series of experiments started in the late 1770s, Galvani explored the nature and effects of "animal electricity", the response to static electricity of nerves in animals. His most famous experiment can be best described in his own words: "*Therefore having noticed that the frog preparations that hung by copper hooks from the iron railings surrounding a balcony of our house contracted not only during thunder storms but also in fine weather, I decided to determine whether or not these contractions were due to the action of atmospheric electricity. ... Finally ... I began to scrape and press the hook fastened to the backbone against the iron railing to see whether by such a procedure contractions might be excited, and whether instead of an alteration in the condition of atmospheric electricity some other changes might be effective. I then noticed frequent contractions, none of which depended on the variations of the weather.*"

In 1780 Galvani constructed a crude electric cell with two different metals and the natural fluids from a dissected frog. These experiments blurred the line between biology, chemistry and electricity and inspired other scientists to do research in the field. Through his experiments, Galvani hit upon the central idea of galvanism: the production of electrical current from the contact of two different metals in a moist environment. However, Galvani interpreted his experiments as proof that animals contain in their nerves and muscles a fluid analogous to ordinary electricity.

Alessandro Volta was among the first to support the new theory but by 1793 he had changed his mind. The Galvani-Volta controversy gave rise to a lively debate among scientists regarding the valid explanation of the phenomenon. Posterity records that it was Volta who was correct.

During the last few years of his life Galvani suffered several personal misfortunes. In 1797 he was stripped of his post at the University of Bologna. A year later, he died, a poor and broken man.

10
Potentiometry

A clean strip of reddish-brown copper metal is immersed in a colourless solution *X* for a week. It is then found to have become silvery-white in colour. The strip is removed from the solution, cleaned thoroughly with distilled water and then heated in a strong flame. The metal strip turns golden-yellow. What is the solution *X*?

The answer to this puzzle can be found at the end of the chapter but to understand it, we will have to learn all about metals and their interaction with ions in solution.

10.1 The Equilibrium between a Metal and Metal Ions in Solution

Consider a metal rod dipped in a solution containing the corresponding metal ions. The metal rod is generally called an *electrode*. A metal electrode dipped in an aqueous solution of its ions is called a *half-cell*. Two possible interactions can take place between the metal atoms on the surface of the electrode and the metal ions surrounding the electrode.

(a) A metal atom on the surface of the electrode may lose *n* electrons and enter the solution as an M^{n+} ion. The electrons left behind on the metal rod cause the metal rod to develop a negative charge. The region of the solution very close to the rod undergoes an increase in positive charge owing to the extra M^{n+} ion.

(b) A metal ion M^{n+} in solution accepts *n* electrons from the metal rod and gets converted to a metal atom *M* on the surface of the rod. Removal of electrons from the surface of the metal rod causes it to acquire positive charge while removal of a positive ion from the solution close to the rod causes the solution to acquire a negative charge.

Given enough time, these two opposing reactions cause equilibrium to be set up between the metal atoms on the rod and the metal ions in solution.

$$M(s) \longleftrightarrow M^{n+}(aq) + ne^-$$

When the metal atoms have a greater tendency to give up electrons and form positive ions in solution, the equilibrium lies more to the right. The loss of electrons is called *oxidation*. Thus the metal atoms get *oxidised* to metal ions.

$$M(s) \rightarrow M^{n+}(aq) + ne^- \text{ (oxidation)} \qquad (10.1)$$

When the metal ions have a greater tendency to acquire electrons and form metal atoms, the equilibrium lies more to the left. The gain of electrons is called *reduction*. Thus the metal ions get *reduced* to metal atoms.

$$M^{n+}(aq) + ne^- \rightarrow M(s) \quad \text{(reduction)} \tag{10.2}$$

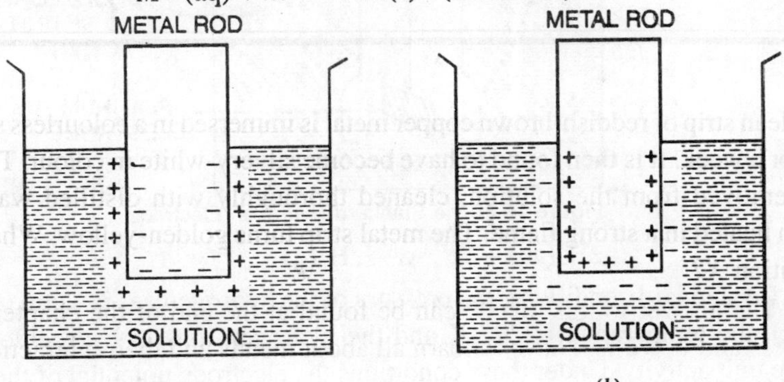

(*a*) (*b*)

Figure 10.1: The origin of Helmholtz double layers around a metal rod dipping into a solution of its ions.

In both cases, an electrical double layer is set-up, as shown in Figure 10.1. Such a layer is called a *Helmholtz double layer*. The double layer is nothing but a separation of charges. Whenever such a separation of charges occurs, a potential difference is also present. In this case, the potential difference is between the metal rod and the surrounding solution. The magnitude of the potential difference developed depends on the concentration of metal ions and the temperature, besides the nature of the metal itself. To measure this potential difference we can think of taking a voltmeter and two pieces of wire. One wire we connect to the rod and the other we dip into the solution. However, this scheme has a flaw. As soon as another metal wire is dipped into the solution, another Helmholtz double layer is set-up between this wire and the metal ions in solution. At best the potential difference between these two Helmholtz double layers can be measured; it is impossible to measure directly the potential difference of a single Helmholtz double layer. To overcome this limitation, it was decided that the electrode potential of the Standard Hydrogen Electrode should be arbitrarily taken as zero. The potential difference between the Standard Hydrogen Electrode and any other electrode would directly give the electrode potential of the other electrode.

10.2 The Standard Hydrogen Electrode

The Standard Hydrogen Electrode (often abbreviated to S. H. E.) consists of hydrogen gas bubbling over a platinum electrode immersed in a solution of hydrochloric acid (Figure 10.2). The platinum electrode is usually coated with finely divided platinum called platinum black. This acts as a catalyst to allow equilibrium between hydrogen gas and the hydrogen ions in solution to be established quickly.

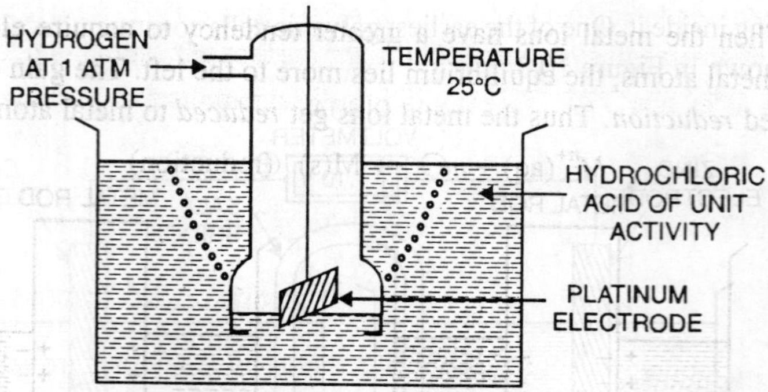

Figure 10.2: The Standard Hydrogen Electrode

$$2H^+(aq) + 2e^- \longleftrightarrow H_2(g)$$

At standard conditions, the pressure of the hydrogen gas is 1 atm (101.325 kPa), the temperature is 298 K and the hydrogen ions in hydrochloric acid have unit activity. Under these conditions the electrode potential of the S.H.E. is arbitrarily taken to be exactly zero volts. Note that the actual electrode potential is not zero but has been defined to be so in order to overcome the limitation that the potential of a single half-cell cannot be measured directly.

10.3 Standard Electrode Potentials

The electrode potential of any metal electrode can be determined by measuring the potential difference between the metal electrode and the Standard Hydrogen Electrode. At standard conditions, the metal ions in solution must have unit activity and the temperature should be 298 K. Under these conditions, the electrode potential measured is known as the *standard electrode potential*. Note that the standard electrode potential always refers to the *reduction* reaction involved. For example, if the standard electrode potential of the zinc electrode is – 0.76 V it implies that the electrode potential of the half-cell reaction

$$Zn^{2+}(aq) + 2e^- \longleftrightarrow Zn(s)$$

is – 0.76 V under standard conditions.

The more negative the value of the standard electrode potential, the greater is the tendency for a metal to give up its electrons, and the stronger is its reducing power. Conversely, the more positive the value of the standard electrode potential, the greater is the tendency for a metal to gain electrons, and the stronger is its oxidising power.

Table 10.1 in Appendix 10 lists out the standard electrode potentials of a number of half-cells.

10.4 Galvanic Cells

A galvanic cell is created when two half-cells are combined in such a way that current flows through an external circuit due to the spontaneous reaction occurring inside it. One of the earliest galvanic cells constructed was the Daniell cell, shown in Figure 10.3.

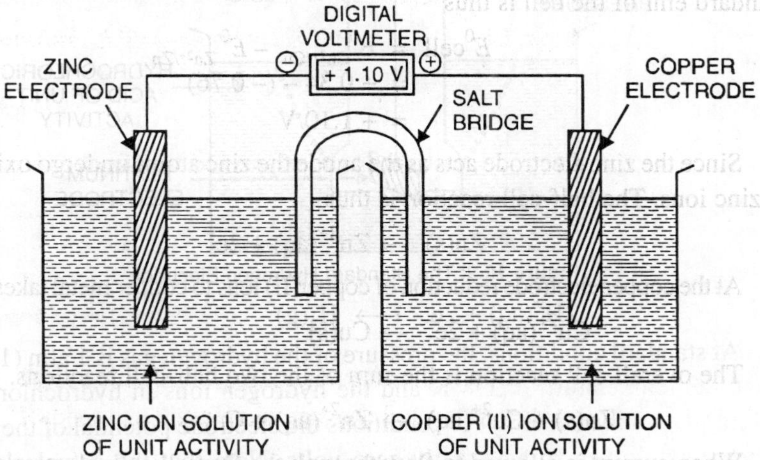

Figure 10.3: The Daniell cell.

One of the half-cells consists of a zinc rod dipping in zinc sulphate solution in which the zinc ions have unit activity. The other half-cell is a copper rod dipping in copper(II) sulphate solution in which copper ions have unit activity. The cells are linked by a *salt bridge*. A typical salt bridge is made by dissolving a salt, such as potassium chloride, in agar. The warm agar is filled in a U-tube. When it cools it sets to a gel. The salt bridge allows charge to flow from one half-cell to another by means of ions. The circuit is completed externally by connecting the zinc electrode to the negative terminal of a voltmeter and the copper electrode to its positive terminal. The voltmeter then shows a reading of + 1.10 V. This reading reflects the difference in the standard electrode potentials of the two electrodes when no current is being drawn from the cell, and is called the *standard electromotive force (emf)* of the cell. The International Union for Pure and Applied Chemistry (IUPAC) recommends that the term 'standard electromotive force' be replaced by the term *zero-current cell potential* but we will continue to use the former because it is more widely used and understood.

It is possible to calculate the standard emf of a combination of any two half-cells. The first step is to determine which half-cell acts as the anode and which as the cathode. The *anode* is the electrode at which *oxidation* takes place while the *cathode* is the electrode at which *reduction* takes place. The electrode with the more negative value of the standard electrode potential acts as the anode. The standard emf of the cell is the standard electrode potential of the cathode minus that of the anode.

$$E^0_{\text{cell}} = E^0_{\text{cathode}} - E^0_{\text{anode}} \quad (10.3)$$

For example, in the case of the Daniell cell, the standard electrode potential of the zinc electrode is -0.76 V while that of the copper electrode is $+0.34$ V.

Since the standard electrode potential of the zinc electrode is more negative, it is chosen as the anode. The copper electrode then acts as the cathode. The standard emf of the cell is thus

$$\begin{aligned} E^0\text{cell} &= E^0_{Cu^{2+}/Cu} - E^0_{Zn^{2+}/Zn} \\ &= +0.34 - (-0.76) \\ &= +1.10 \text{ V} \end{aligned}$$

Since the zinc electrode acts as the anode the zinc atoms undergo oxidation to zinc ions. The half-cell reaction is thus

$$Zn(s) \rightarrow Zn^{2+}(aq) + 2e^-$$

At the copper electrode reduction of copper(II) ions to copper metal takes place.

$$Cu^{2+}(aq) + 2e^- \rightarrow Cu(s)$$

The overall cell reaction is the sum of the two half-cell reactions.

$$Zn(s) + Cu^{2+}(aq) \rightarrow Zn^{2+}(aq) + Cu(s)$$

When current is allowed to flow through the Daniell cell, electrons move from the zinc electrode through the external circuit to the copper electrode. The electrical energy supplied to the cell can be used to perform useful work. As the current continues to flow the voltage of the cell keeps dropping till it falls to zero. At this point, there is no more potential difference between the two half-cells and the reaction has reached equilibrium.

$$Zn(s) + Cu^{2+}(aq) \longleftrightarrow Zn^{2+}(aq) + Cu(s)$$

Galvanic ells are used as a source of stored electrical energy. When two or more cells are connected together they form a battery.

10.5 Cell Notations

Cell notations are used to provide a quick and efficient way to describe a cell.

For example, consider the Daniell cell. It can be written as

$$Zn(s) \mid Zn^{2+}(aq),\ a_{Zn^{2+}} = 1 \mid\mid Cu^{2+}(aq),\ a_{Cu^{2+}} = 1 \mid Cu(s)$$

The electrode that acts as anode is written first, followed by the cathode.

A single vertical line, |, denotes a phase boundary between the two chemical species separated by the vertical line. A double vertical line, ||, denotes the presence of a salt bridge between the two half-cells. Sometimes the formula of the salt used to make the salt bridge is written in between the two half-cells. This may be necessary in case potassium chloride, the salt normally used, interferes in the cell reaction. For example, a cell formed from silver and copper half-cells would require potassium nitrate to form the salt bridge since chloride ions from potassium chloride would cause silver ions to precipitate. The cell notation for such a cell would be

$$Cu(s) \mid Cu^{2+}(aq) \mid KNO_3 \mid Ag^+(aq) \mid Ag(s)$$

10.6 The Nernst Equation

Consider a general reaction of the type

$$aA + bB \rightarrow cC + dD$$

The reaction quotient, Q, of this reaction is defined to be

$$Q = \frac{(a_C)^c\,(a_D)^d}{(a_A)^\alpha\,(a_B)^b} \tag{10.4}$$

In general, a_J represents the activity of the Jth species.

The Gibbs free energy of this reaction $\Delta_r G$ is given by

$$\Delta_r G = \Delta_r G^0 + RT\,ln\,Q \tag{10.5}$$

where $\Delta_r G^0$ is the standard Gibbs energy of the reaction
R is the universal gas constant
T is the absolute temperature.

The relation between the Gibbs energy of the cell reaction and the emf of the cell is

$$-nFE_{cell} = \Delta_r G \tag{10.6}$$

where n is the number of electrons exchanged in the redox reaction
F is the Faraday constant

This equation is the key equation relating electrical measurements to the thermodynamic properties of the system.

Substituting equation (10.6) in equation (10.5) we get

$$-nFE_{cell} = -nFE^0_{\,cell} + RT\,ln\,Q \tag{10.7}$$

On rearranging this equation we get

$$E_{cell} = E^0_{\,cell} - \frac{RT}{nF}\,ln\,Q \tag{10.8}$$

where $E^0_{\,cell}$ is the standard emf of the cell.

Equation (10.8) is known as the *Nernst equation*. It predicts that the emf of a cell depends on the concentration of ions taken.

For example, consider the Daniell cell in which the zinc ion concentration is kept fixed at 1 mol dm^{-3} while the copper(II) ion concentration is varied between 1 mol dm^{-3} and 10^{-6} mol dm^{-3}. Since the potential of the copper half-cell is expected to vary with change in copper(II) ion concentration, the emf of the Daniell cell will also show corresponding variation. Table 10.2 shows some typical results.

Table 10.2: Variation in the emf of a Daniell cell with variation in copper(II) ion concentration

Concentration of Cu^{2+}(aq)/(mol dm^{-3})	Ecell / (V)	$E_{Cu2+/Cu}$ / (V)
1.0	1.1	0.340
10^{-1}	1.070	0.310
10^{-2}	1.041	0.281
10^{-3}	1.011	0.251
10^{-4}	0.982	0.222
10^{-5}	0.953	0.193
10^{-6}	0.923	0.163

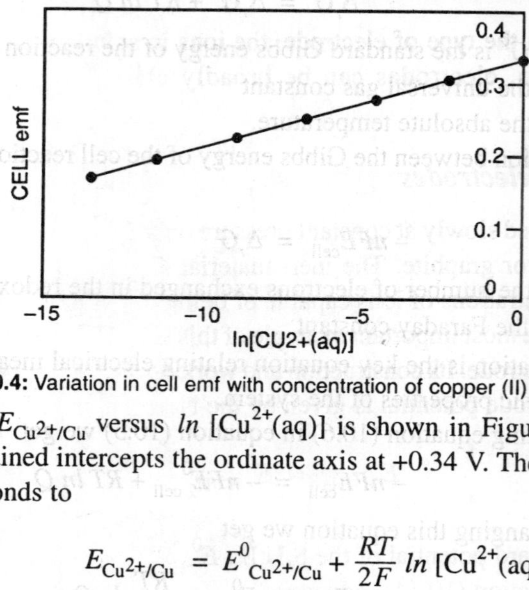

Figure 10.4: Variation in cell emf with concentration of copper (II) ions.

A graph of $E_{Cu^{2+}/Cu}$ versus ln [Cu^{2+}(aq)] is shown in Figure 10.4. The straight-line obtained intercepts the ordinate axis at +0.34 V. The equation of this line corresponds to

$$E_{Cu^{2+}/Cu} = E^0_{Cu^{2+}/Cu} + \frac{RT}{2F} ln [Cu^{2+}(aq)] \qquad (10.9)$$

For a general reaction

$$\text{oxidised form} + ne^- \rightarrow \text{reduced form} \qquad (10.10)$$

the Nernst equation can be written as

$$E_{ox/red} = E^0_{ox/red} - \frac{RT}{nF} \ln \frac{[\text{reduced form}]}{[\text{oxidised form}]} \qquad (10.11)$$

A special case of the Nernst equation allows us to relate the standard emf of the cell with the equilibrium constant of the reaction. When the cell reaction has reached equilibrium the cell emf becomes zero. At this stage, the reaction quotient, Q, is equivalent to the equilibrium constant, K, of the reaction.

Substituting $Q = K$ and $E_{cell} = 0$ into equation (10.8),

$$0 = E^0_{cell} - \frac{RT}{nF} ln K \qquad (10.12)$$

Rearranging this equation, we get

$$ln\,K = \frac{nF}{RT}E^0_{cell} \qquad (10.13)$$

For example, in the case of the Daniell cell E^0_{cell} is +1.1 V and n = 2. Substituting these values in equation (10.13) we get $K = 1.5 \times 10^{37}$. The high value of K indicates that the forward reaction

$$Cu^{2+} + Zn(s) \longleftrightarrow Cu(s) + Zn^{2+}(aq)$$

goes virtually to completion.

10.7 Types of Electrodes

Depending on the type of electrode, the ions in solution, and the interaction between them, electrodes can be broadly classified into the following categories.

10.7.1 Gas electrodes

A gas is bubbled slowly at constant pressure on the surface of an inert material like platinum or graphite. The inert material dips into a solution containing known concentrations of ions capable of being oxidised or reduced to the gas reversibly. The most important example of this type of electrode is the Standard Hydrogen Electrode. Its construction and working has been described in Section 10.2. Its electrode potential is given by the Nernst equation as

$$E_{H^+/H_2} = E^0_{H^+/H_2} - \frac{RT}{2F}\ln\frac{[1]}{[a_{H^+}]^2} \qquad (10.14)$$

The standard potential of the S.H.E., $E^0_{H^+/H_2}$, is chosen to be zero volts. Therefore equation (10.14) reduces to

$$E_{H^+/H_2} = -\frac{RT}{2F}\ln\frac{[1]}{[a_{H^+}]^2}$$

$$E_{H^+/H_2} = \frac{RT}{F}\ln\frac{[1]}{[a_{H^+}]^2} \qquad (10.15)$$

The last term in this equation, $ln\,[\,1/a_{H^+}]$, can be expressed in terms of common logarithms as follows

$$E_{H^+/H_2} = -2.303\frac{RT}{F}\log_{10}\frac{[1]}{[a_{H^+}]} \qquad (10.16)$$

The concept of pH was introduced in 1909 by S.P.L. Sorenson, a Danish biochemist. Today it is mathematically defined as

$$pH = \log_{10}\frac{[1]}{[a_{H^+}]} \qquad (10.17a)$$

or
$$pH = -\log_{10}[a_{H^+}] \tag{10.17b}$$

The electrode potential can thus be expressed in terms of pH by substituting equation (10.17a) in equation (10.16).

$$E_{H^+/H_2} = -2.303\frac{RT}{F}\text{pH} \tag{10.18a}$$

At 298 K, this is the same as

$$E_{H^+/H_2} = -0.0592\text{ pH} \tag{10.18b}$$

The practical utility of the S.H.E. is limited due to the several disadvantages. The solution should not contain any substances that may be oxidised or reduced by hydrogen. High molecular weight compounds like proteins should be absent because they are adsorbed on the platinum and interfere seriously with the functioning of the electrode. Solutions that contain volatile acids like carbon dioxide, or volatile alkalis like ammonia cannot be determined accurately because they are swept out of the solution by the hydrogen gas.

In view of all these difficulties the S.H.E. is no longer used regularly in the laboratory. It has been replaced by several other reference electrodes like the calomel electrode, silver-silver chloride electrode, etc. However, it remains the only electrode that can accurately measure pH values above 12 since most other pH electrodes fail entirely or are unreliable in very strong alkaline medium.

Another example of the gas electrode is the chlorine electrode, in which chlorine gas is bubbled over graphite in contact with a solution of chloride ions. The electrode reaction is

$$Cl_2(g) + 2e^- \longleftrightarrow 2Cl^-(aq)$$

The electrode potential is given by

$$E_{Cl_2/Cl^-} = E^0_{Cl_2/Cl^-} - \frac{RT}{2F}\ln\frac{a^2_{Cl^-}}{p_{Cl_2}} \tag{10.19}$$

where p_{Cl_2} is the partial pressure of the chlorine gas.

10.7.2 Metal-metal ion electrodes

A metal rod dips into a solution containing its ions. Equilibrium is established between the metal and the metal ions in solution.

$$M^{n+}(aq) + ne^- \rightarrow M(s)$$

The Nernst equation for this electrode reaction is

$$E_{M^{n+}/M} = E^0_{M^{n+}/M} - \frac{RT}{nF}\ln\frac{[1]}{[a_{M^{n+}}]} \tag{10.20}$$

An example is the Cu^{2+}/Cu electrode that consists of a copper electrode dipped in a solution of copper(II) ions. Metals that react with the solvent cannot be used in this type of electrode. Alkali and alkaline earth metals react with water; zinc reacts with aqueous acidic solutions. Certain metals are easily oxidised by oxygen, which must be therefore removed from the solution by bubbling nitrogen gas through it.

10.7.3 Amalgam electrodes

A metal dissolves in liquid mercury to form an amalgam. Active metals like the alkali and alkaline earth metals, which react with water, can be made reversible to their ions by dissolving them in mercury. The electrical contact is made by a platinum wire dipping in the amalgam. The mercury does not participate in the electrode reaction, which is

$$M^{n+} + ne^- \rightarrow M(Hg)$$

The electrode is represented as $M(Hg) \mid M^{n+}(aq)$.

Examples of this type of electrode are the sodium amalgam electrode, $Na(Hg) \mid Na^+(aq)$ and the cadmium amalgam electrode, $Cd(Hg) \mid Cd^{2+}(aq)$. The potential of an amalgam electrode depends on the activity of metal ions in solution as well as the activity of the metal in mercury.

10.7.4 Redox electrodes

A redox electrode consists of an inert material like platinum or graphite dipping into a solution containing a chemical species in two different oxidation states. The transfer of electrons between the species takes place through the inert material. An example is the iron(III)-iron(II) ion-platinum electrode. The electrode reaction is

$$Fe^{3+} + e^- \longleftrightarrow Fe^{2+}$$

The electrode potential is given by

$$E_{Fe^{3+}/Fe^{2+}} = E^0_{Fe^{3+}/Fe^{2+}} - \frac{RT}{F} \ln \frac{[a_{Fe^{2+}}]}{[a_{Fe^{3+}}]} \qquad (10.21)$$

10.7.4.1 Quinhydrone electrode

Another example of a redox electrode is that of the quinhydrone electrode in which platinum dips into a solution saturated with quinhydrone. Quinhydrone is a slightly soluble compound formed by the combination of one mole of quinone (Q) and one mole of hydroquinone (H_2Q). The electrode reaction is

$$\underset{\text{Quinone}}{C_6H_4O_2} + 2H^+ + 2e^- \longleftrightarrow \underset{\text{Hydroquinone}}{C_6H_4(OH)_2} \qquad (10.22)$$

This electrode is very easy to prepare and handle. Pure solid quinhydrone is dissolved in the solution to be measured until the solution is saturated and excess is present. A platinum wire is dipped in this solution.

The electrode potential is given by

$$E_{Q/H_2Q} = E^0_{Q/H_2Q} - \frac{RT}{2F} \ln \frac{[a_{H_2Q}]}{[a_Q][a_{H^+}]^2} \quad (10.23)$$

Since quinone and hydroquinone are obtained by dissolving quinhydrone in the solution, $a_{H_2Q} = a_Q$. Hence equation (10.23) reduces to

$$E_{Q/H_2Q} = E^0_{Q/H_2Q} - \frac{RT}{2F} \ln \frac{[1]}{[a_{H^+}]^2} \quad (10.24)$$

This further simplifies to

$$E_{Q/H_2Q} = E^0_{Q/H_2Q} - 2.303 \frac{RT}{F} \text{pH} \quad (10.25)$$

The standard potential of the quinhydrone electrode is given by

$$E^0_{Q/H_2Q} = -0.6994 + 7.36 \times 10^{-4} (t - 25) \quad (10.26)$$

where t is the temperature in degree Celsius.

At 298 K the potential of this electrode is given by

$$E = -0.6994 - 0.0592 \text{ pH} \quad (10.27)$$

The quinhydrone electrode cannot be used in solutions that would react with quinone or hydroquinone. Hydroquinone being a weak acid ($K_1 = 10^{-10}$) the electrode cannot be used above pH 8.5 when the dissociation of hydroquinone becomes appreciable. Another drawback is that quinone is oxidised by air in strongly alkaline medium. In spite of all this, the quinhydrone electrode is frequently used in titrations of acids with bases since the end point is reached below pH 9 in most cases.

10.7.5 Metal-insoluble salt electrodes

Metal-insoluble salt electrodes consist of a metal, M, in contact with an insoluble salt of the metal, MX, and a solution containing the anion of the insoluble salt, X^-. The electrode potential depends on the concentration of the anion, X^-. Two common examples of this type of electrode are the silver-silver chloride electrode and the calomel electrode.

10.7.5.1 The calomel electrode: $Hg(l) \mid Hg_2Cl_2(s) \mid Cl^-(aq)$

Mercury metal is in equilibrium with mercury(I) ions in solution.

$$2Hg(l) \longleftrightarrow Hg_2^{2+}(aq) + 2e^-$$

Since mercury(I) chloride is a sparingly soluble salt, the solution is saturated with it, which gives the following equilibrium:

$$Hg_2^{2+}(aq) + 2Cl^-(aq) \longleftrightarrow Hg_2Cl_2(s)$$

The sum of these two reactions gives the net half-cell reaction.

$$2Hg(l) + 2Cl^-(aq) \longleftrightarrow Hg_2Cl_2(s) + 2e^-$$

Depending on the concentration of chloride ions used, there are three types of calomel electrode. Table 10.3 summarises their characteristics.

Table 10.3: Types of calomel electrode
(t is the temperature in degree Celsius)

Concentration of KCl /(N)	*Name given to electrode*	*Electrode potential / (V)*
saturated solution	Saturated Calomel Electrode (S.C.E.)	$+ 0.242 - 7.6 \times 10^{-4} (t - 25)$
1.0	Normal Calomel Electrode (N.C.E.)	$+ 0.280 - 2.4 \times 10^{-4} (t - 25)$
0.1	Decinormal Calomel Electrode	$+ 0.334 - 7 \times 10^{-5} (t - 25)$

The Saturated Calomel Electrode is shown in Figure 10.5(a). A small quantity of pure mercury metal is placed at the bottom of the electrode vessel. A thin layer of calomel paste prepared by grinding together mercury(I) chloride, mercury metal, and some potassium chloride solution is placed over the pure mercury metal. A platinum wire sealed in a thin capillary tube is inserted into the mercury layer from top to make electrical contact. The electrode vessel is filled with saturated potassium chloride solution through the bent side tube. The side tube is kept immersed in a saturated solution of potassium chloride. When using this electrode to construct a galvanic cell, the salt bridge is dipped in this solution. Modern calomel electrodes are of the dip type, as shown in Figure 10.5 (b).

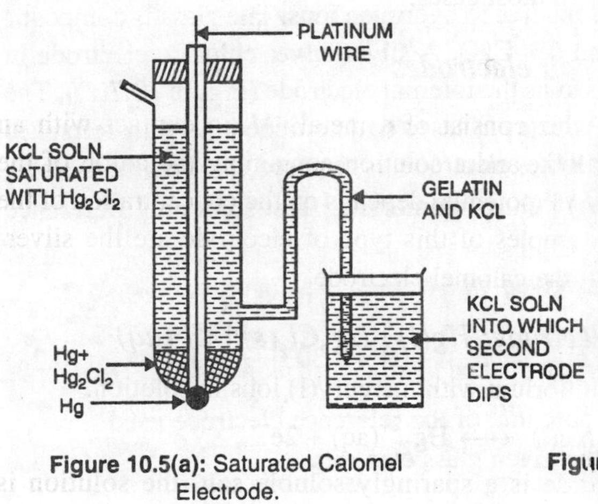

Figure 10.5(a): Saturated Calomel Electrode.

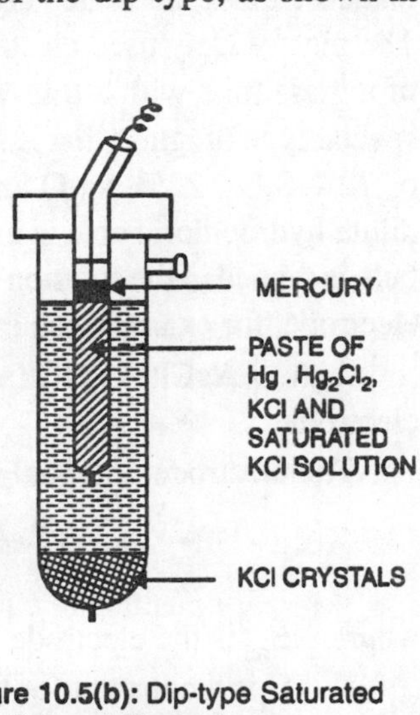

Figure 10.5(b): Dip-type Saturated Calomel Electrode.

10.7.5.2 The silver-silver chloride electrode: Ag(s) | AgCl(s) | Cl⁻(aq)

Silver metal is in equilibrium with silver ions in solution.

$$Ag(s) \longleftrightarrow Ag^+(aq) + e^-$$

Since the solution is saturated with AgCl, the following equilibrium also exists

$$Ag^+(aq) + Cl^-(aq) \longleftrightarrow AgCl(s)$$

The net electrode reaction is the sum of these two reactions:

$$Ag(s) + Cl^-(aq) \longleftrightarrow AgCl(s) + e^-$$

The silver-silver chloride reference electrode is shown in Figure 10.6. The electrode vessel is a tube whose bottom is closed by a sintered glass disk. The lower half of the tube is filled with agar gel in saturated potassium chloride. The upper half contains saturated potassium chloride solution to which a drop or two of 1 M silver nitrate solution has been added to saturate it with silver chloride. A thick wire dips in this solution. The potential of this electrode is +0.197 V.

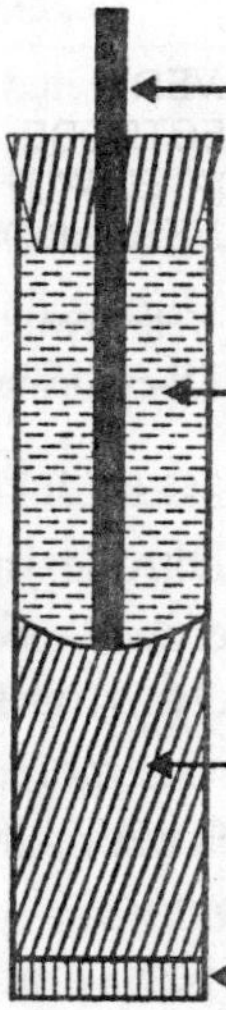

Figure 10.6: The silver/silver chloride electrode.

10.7.6 Ion-selective electrodes

An ion-selective electrode contains a membrane that is sensitive to one particular ion in the solution. The membrane can be made of glass, an ionic crystal or a liquid. The most widely used electrode of this type is the *glass electrode.* It consists of a glass tube with a thin-walled bulb at the bottom. The bulb is made of a special type of glass that is sensitive to hydrogen ions. The glass is composed of 72% SiO_2, 22% Na_2O and 6% CaO. A silver-silver chloride electrode in dilute hydrochloric acid is used as the internal electrode (Figure 10.7(a)). The bulb is dipped in the solution whose pH has to be determined. Another reference electrode, for example the calomel electrode, is required to complete the cell.

Ag(s) | AgCl(s) | HCl(aq) | Glass membrane | Test solution || Reference electrode

The electrode potential is given by

$$E_{cell} = E_{ref} - (G - 2.303 \frac{RT}{F} \text{pH}) \qquad (10.28)$$

where E_{ref} is the electrode potential of the reference electrode used
G is a constant for the given glass electrode.

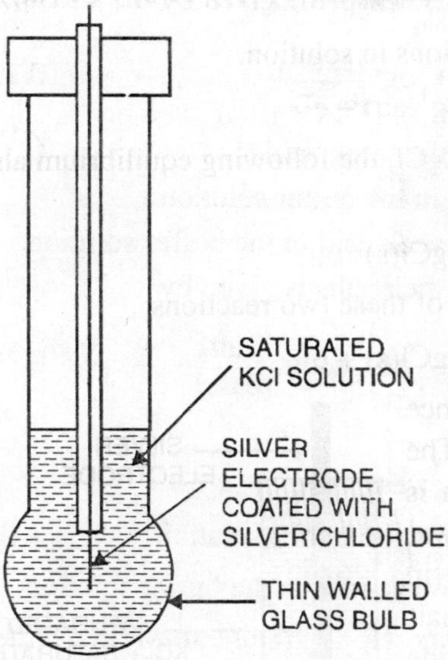

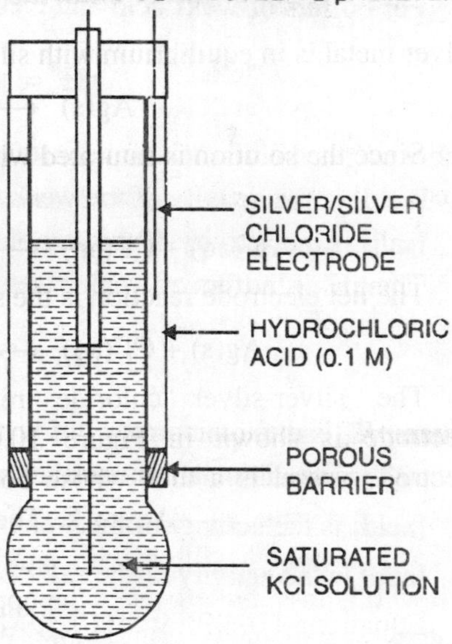

Figure 10.7(a): The glass electrode. **Figure 10.7(b):** The combination glass electrode.

Modern glass electrodes are combination electrodes in which the glass electrode and the reference electrode are arranged in one piece of equipment, as shown in Figure 10.7(b). The emf of the cell can be measured using a digital voltmeter but the more common practice is to convert the emf into a pH reading by using a pH meter.

The glass electrode can be used in the presence of strong oxidising or reducing agents, in viscous or semi-solid media, and in the presence of high molecular weight substances like proteins, which interfere seriously with other electrodes. The presence of water in the glass membrane is very important for the functioning of the glass electrode. The membrane should not be allowed to dry out under any circumstances. The glass electrode shows an alkaline error above pH 9 and hence the observed pH is larger than the actual value.

10.8 Buffer Solutions

When a solution of a definite pH is required, it is not enough to add acid or alkali to distilled water till the desired pH is reached, because the pH of the resulting solution is unstable. Instead, *buffer solutions* are used. These are solutions whose pH is not altered to any great extent by the addition of small quantities of either acid or base. The pH remains stable due to the presence of a chemical equilibrium that shifts in the correct direction to maintain the pH of the solution. Buffer solutions are generally composed of a weak acid and its salt with a strong base (e.g., acetic acid and sodium acetate) or a weak base and its salt with a strong base (e.g., ammonium hydroxide and ammonium chloride).

For buffers of weak acid and their salts, the pH is given by the equation

$$\text{pH} = pK_a + \log \frac{[\text{salt}]}{[\text{acid}]} \tag{10.29}$$

where pK_a is the dissociation constant of the weak acid

[acid] is the activity of the weak acid in the buffer solution

[salt] is the activity of the salt of the weak acid in the buffer solution.

The pH of buffers of weak bases and their salts is given by

$$\text{pH} = pK_w - pK_b + \log \frac{[\text{salt}]}{[\text{base}]} \tag{10.30}$$

where pK_w is the ionic product of water

pK_b is the dissociation constant of the weak base

[acid] is the activity of the weak base in the buffer solution

[salt] is the activity of the salt of the weak base in the buffer solution.

Equations (10.29) and (10.30) are known as *Henderson-Hasselbalch equations* for buffers. It is obvious from these two equations that when [salt] = [acid], pH = pKa in acid buffers and when [salt] = [base], pH = pKw in basic buffers.

Table 10.4 in Appendix 10 lists some commonly used buffer solutions and their pH values.

10.9 Potentiometric Titrations

The Nernst equation states that the potential of an electrode depends on the concentration of the ions that are reversible to the electrode and are in contact with it. If the concentration of these ions is changed either by neutralisation, precipitation or a redox reaction, the electrode potential will also change. Potentiometric titrations are based on this principle. They are particularly useful for dilute solutions and reactions for which no colour indicator is available. The electrode in contact with the solution whose ions are to be titrated is called the *indicator electrode.*

10.9.1 End point determination

The end point in a potentiometric titration coincides with a very rapid change in the potential of the indicator electrode. This rate of emf change is usually much greater than at any other point in the course of the titration. The end point can thus be accurately and precisely determined from the change in emf versus titration volume. The end point determination is done graphically. The titrant is added in very small increments. The titration curve is plotted simultaneously point by point. The cell emf is plotted on the ordinate and the corresponding volume of the titrant added on the abscissa. The cell emf varies gradually initially but as the end point approaches, its rate of change is more rapid. The titrant should be added in very small increments near the end point to obtain a large number of corresponding points on the titration curve, which will help in obtaining an accurate inflexion point. After the end point has been passed, the rate of change of cell emf comes down again. The resulting titration curve is shown in Figure 10.8a.

The end point can be precisely determined from the first and second derivative curves of the titration curve. A first derivative curve is shown in Figure 10.8b. The end point occurs at the titration volume corresponding to the maxima of the curve. The second derivative curve is shown in Figure 10.8c. The end point is found at the point where the second derivative curve becomes zero, i.e., where the second derivative curve intercepts the abscissa.

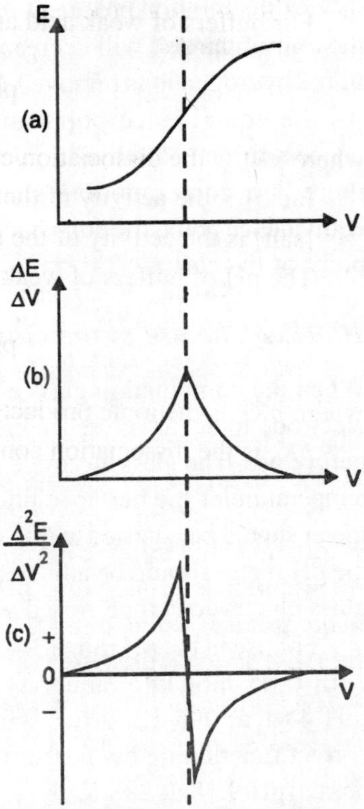

Figure 10.8: The graphical determination of end point in a potentiometric titration.

10.9.2 Acid-base potentiometric titrations

Acid-base titrations involve a change in pH over the course of the titration. Therefore they are also known as pH-metric titrations. They can be carried out with the quinhydrone electrode or the glass electrode serving as the indicator electrode. The Saturated Calomel electrode generally serves as the reference electrode.

10.9.2.1 The use of the quinhydrone electrode

When the quinhydrone electrode is used, the acid solution must be taken in the beaker and the alkali in the burette because quinhydrone is unstable in alkaline solutions. The following cell is set-up.

S.C.E. || analyte solution saturated with quinhydrone | Pt(s)

The emf of the cell is given by

$$E_{cell} = E_{Q/H_2Q} - E_{S.C.E.}$$

$$E_{cell} = E^0_{Q/H_2Q} - 2.303\frac{RT}{F}\,\text{pH} - E_{S.C.E.} \quad (10.31)$$

As the titration proceeds, the pH of the solution will increase and therefore the emf of the cell will decrease. The decrease in emf depends on the fraction of the hydrogen ions removed. The cell emf therefore decreases slowly at first. As the equivalence point approaches, the fraction of the hydrogen ion concentration neutralised by each drop of alkali solution increases rapidly and there is a corresponding sharp change in the emf of the cell. After the equivalence point, the emf changes slowly. The end point occurs at the inflexion point of the curve.

10.9.2.2 The use of the glass electrode

When the combination glass electrode is used, there is no need for any other electrode to act as reference electrode. The pH meter should be *standardised* before use. The glass electrode is dipped in a buffer solution of known pH. The temperature of the buffer solution is measured. The temperature dial of the pH meter should be adjusted to this value. After about half an hour the pH reading on the pH meter should be adjusted to match that of the buffer solution used. The glass electrode is then rinsed with distilled water. It is now ready for use.

For use in the pH range 3 – 10, the glass electrode is generally standardised with 0.05 mol kg^{-1} aqueous solution of potassium hydrogen phthalate (pH 4.01 at 298 K). After standardising, the pH meter can be checked for correct functioning by measuring the pH of 0.01 mol kg^{-1} aqueous solution of borax (pH 9.18 at 298 K).

Table 10.5 in Appendix 10 lists out the pH values and compositions of several buffer solutions used to standardise pH meters.

10.9.2.3 Titration of a monobasic acid with a strong base

A known volume of a weak acid like acetic acid is taken in the titration vessel and a standard solution of a strong base like NaOH is taken in the burette. The titration is carried a little past the end point so that a complete titration curve is obtained. The end point is located by drawing the first derivative graph of $\Delta pH/\Delta V$ versus V. The pH where the volume of the titrant added is half that required for neutralisation corresponds to the pK_a of the weak acid.

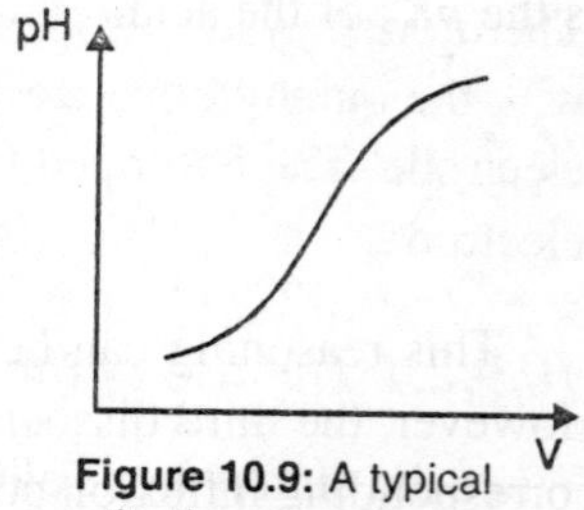

Figure 10.9: A typical titration curve for a monobasic acid.

$$pK_a = \text{pH for which } V = V_{\text{end point}}/2 \qquad (10.32)$$

Figure 10.9 shows a typical plot of pH versus volume of alkali added for a monobasic acid.

10.9.2.4 Titration of a polybasic acid with a strong base

In the case of polybasic acids there is more than one ionisable hydrogen ion. The hydrogen ions are produced by the stepwise dissociation of the acid, each step having its own dissociation constant K_1, K_2, etc. During a potentiometric titration, each step will give rise to an inflexion point on the titration curve.

Consider the case of a *dibasic acid.* When the first dissociation constant, K_1, is at least a hundred times greater than the second, K_2, the two inflexion points can be clearly distinguished on the titration curve. When the ratio K_1/K_2 is less than hundred, the two inflexion points overlap, and only the inflexion point corresponding to the second dissociation step appears on the titration curve. In this case, the pK_1 and pK_2 values of the acid cannot be determined graphically.

A typical titration curve of a weak dibasic acid with K_1/K_2 greater than hundred is shown in Figure 10.10. Let V_1 be the volume corresponding to the first end point, and V_2 the volume corresponding to the second. The first dissociation constant is determined the same way as for a monobasic acid. The pH corresponding to the addition of one and a half equivalents of the titrant is the pK_2 of the acid.

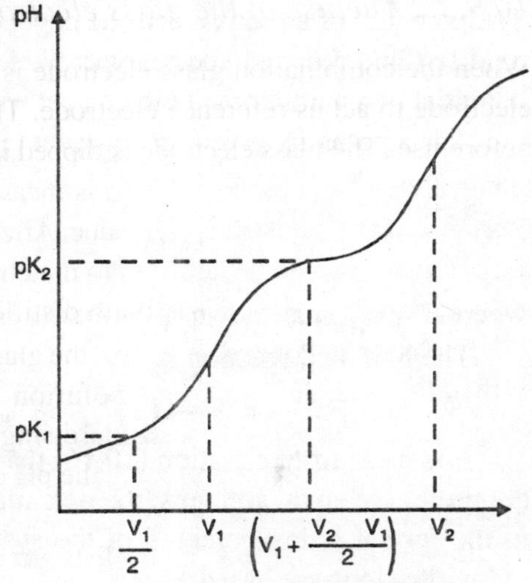

Figure 10.10: A typical titration curve for a weak dibasic acid with $K_1/K_2 > 100$.

$$pK_1 = \text{pH for which } V = V_1/2$$

$$pK_2 = \text{pH for which } V = \frac{(V_2 - V_1 + V_1)}{2} \qquad (10.32)$$

This reasoning can be extended to a *tribasic acid* like phosphoric acid. However, the third dissociation step of phosphoric acid is very weak and the corresponding inflexion point is not sharp enough to determine accurately. To overcome this difficulty, the pH of the solution is lowered after the second end point and the titration resumed from there. Since the end point determination does not depend on the absolute values of pH but only on the rate of change of pH with titrant volume, this method works well. Solid calcium chloride is added to the solution. It reacts with disodium hydrogen phosphate to liberate hydrochloric acid. The liberation of a strong acid causes the pH of the solution to fall.

$$2Na_2HPO_4 + 3CaCl_2 \rightarrow 2HCl + Ca_3(PO_4)_2\downarrow + 4NaCl$$

The quantity of hydrochloric acid liberated is equivalent to the acid content of disodium hydrogen phosphate. The third dissociation constant can be calculated from the formula

$$pK_3 = 2(\text{pH at } V_2) - pK_2 \tag{10.33}$$

The dissociation constants of a number of acids can be found in Table 10.6 in Appendix 10.

10.9.3 Redox titrations

The potential of a redox electrode depends on the ratio of the activities of the oxidised and reduced species with which it is in contact. For example, the potential of a platinum wire in contact with a solution of iron (III) and iron(II) ions has a potential given by equation (10.16), replicated here:

$$E_{Fe^{3+}/Fe^{2+}} = E^0_{Fe^{3+}/Fe^{2+}} - \frac{RT}{F} ln \frac{[a_{Fe^{2+}}]}{[a_{Fe^{3+}}]} \tag{10.16}$$

where $E^0_{Fe^{3+}/Fe^{2+}}$ is known as the *formal redox potential* of the Fe^{3+}/Fe^{2+} system.

The half–cell reaction is

$$Fe^{3+} + e^- \longleftrightarrow Fe^{2+} \qquad E^0_{Fe^{3+}/Fe^{2+}} = 0.771 \text{ V}$$

It is clear from equation (10.16) that when the iron(III) and iron(II) ions are present in equal amounts, the potential of the electrode will become equal to the formal redox potential of the system. This is the principle on which redox titrations are based.

An example is the titration of an iron(II) salt with a cerium(IV) salt. The reaction is

$$Fe^{3+} + Ce^{4+} \longleftrightarrow Fe^{2+} + Ce^{3+}$$

A known volume of an iron(II) salt solution is taken in the titration vessel and a platinum electrode is dipped in it. This half-cell is coupled with a reference electrode like the Saturated Calomel Electrode. The cell can be represented as

$$\text{S.C.E.} \parallel Fe^{3+}(aq),\ Fe^{2+}(aq) \mid Pt(s)$$

The emf of the cell is given by the equation

$$E_{cell} = E_{Fe^{3+}/Fe^{2+}} - E_{S.C.E.}$$

$$E_{cell} = E^0_{Fe3+/Fe2+} - \frac{RT}{F} ln \frac{[a_{Fe^{2+}}]}{[a_{Fe^{3+}}]} - E_{S.C.E.} \tag{10.34}$$

Initially there are only iron(II) ions in the solution, making the potential of the platinum electrode practically zero. As cerium(IV) ions are added to the solution, iron(III) ions are formed. The redox couple formed causes the electrode potential to increase gradually. At the end point there is a very sharp increase in the electrode potential. The end point of the titration is determined graphically from the first and second derivative curves. At the end point, all the iron(II) ions present have been oxidised to iron(III) ions. Therefore at a point halfway to the end point, the iron(II) and iron(III) ions should be present in equal amounts. The cell emf corresponding to half the volume of titrant required for end point, $(E_{cell})_{e.p./2}$, is determined.

$$(E_{cell})_{e.p./2} = E^0_{Fe^{3+}/Fe^{2+}} - E_{S.C.E.}$$

$$E^0_{Fe^{3+}/Fe^{2+}} = (E_{cell})_{e.p./2} + E_{S.C.E.}$$

A similar technique is employed for determining the formal redox potential of the Ce^{4+}/Ce^{3+} system.

$$Ce^{4+} + e^- \rightarrow Ce^{3+} \qquad E^0_{Ce^{4+}/Ce^{3+}} = 1.61\ V$$

Cerium(IV) ions are taken in the titration vessel and iron(II) ions are added from the burette. The cell can be represented as

$$S.C.E. \parallel Ce^{4+}(aq),\ Ce^{3+}(aq)\ |\ Pt(s)$$

The emf of the cell is given by the equation

$$E_{cell} = E_{Ce^{4+}/Ce^{3+}} - E_{S.C.E.}$$

$$E_{cell} = E^0_{Ce^{4+}/Ce^{3+}} - \frac{RT}{F} ln \frac{[a_{Ce^{3+}}]}{[a_{Ce^{4+}}]} - E_{S.C.E.} \quad (10.35)$$

During the titration, the cell emf drops slowly at first and very sharply at the end point. At the end point,

$$E^0_{Ce^{4+}/Ce^{3+}} = (E_{cell})_{e.p./2} + E_{S.C.E.}$$

In a redox potentiometric titration the formal redox potentials of the two redox systems should be very different in order to obtain a sharp change in the emf at the end point.

10.9.4 Precipitation titrations

In precipitation titrations, an ion reversible with the electrode is removed from the solution by the titrant as a precipitate. The change in concentration of the ion to which it is reversible causes a change in the potential of the electrode. Precipitation titrations are based on this principle.

The most important examples of this type of titration are those involving sparingly soluble salts of silver, mercury and lead. Titrations involving silver are often called *argentimetric titrations*. Anions that can be determined are the halide ions, thiocyanate ion, cyanide ion, sulphate ion, sulphide ion, ferrocyanide ion, and anions of various organic acids. The indicator electrode may be either the metal involved in the reaction (e.g., silver or mercury electrodes for titrations involving silver or mercury ions), or an electrode whose potential is determined by the anion in question (e.g., a $Pt/I_2, I^-$ electrode used for titrations involving iodide ions).

The general reaction in a precipitation reaction is

$$M^{n+}(aq) + X^{y-}(aq) \longleftrightarrow M_yX_n(s)$$

Mechanical stirring is very important in precipitation titrations in order to speed up the rate at which equilibrium is established between the ions in solution and the precipitate. The equilibrium constant of the reaction is the reciprocal of the solubility product of the sparingly soluble salt, M_yX_n. The larger the value of the equilibrium constant the greater will be the rate of change of cell emf at the end point. Thus compounds with smaller values of solubility product give a sharper end point. Precipitates, which are composed of 1:1 molar ratio of ions, give rise to titration curves that are symmetrical about the end point. The maxima of the first derivative curve $\Delta E/\Delta V$ coincides with the end point of the titration.

Table 10.7 in Appendix 10 summarises the solubility product data of a number of silver, mercury and lead salts commonly used in precipitation titrations.

A classic example of a precipitation titration is the argentimetric titration of a mixture of iodide, bromide and chloride ions. A silver wire dipping in this mixture serves as the indicator electrode. Its potential is given by

$$E_{Ag+/Ag} = E^0_{Ag+/Ag} - \frac{RT}{F} ln \frac{[1]}{[a_{Ag^+}]} \qquad (10.36)$$

The titration reaction is essentially

$$Ag^+(aq) + X^-(aq) \rightarrow AgX(s)$$

where X^- represents the halide ion.

Since silver(I) iodide has the smallest solubility product among the three silver halides, it will be precipitated first. The precipitation continues until the silver ion concentration is equal to $(K_{sp})_{AgBr}/[Br^-]$. At this point, more than 99% of the iodide ions have been removed from the solution. The titration curve shows a very rapid change in cell emf at this point. Further addition of silver ions causes silver(I) bromide to precipitate until the silver ion concentration reaches a value $(K_{sp})_{AgCl}/[Cl^-]$. The titration curve shows another rapid change in cell emf. Beyond this point, precipitation of silver(I) chloride commences. Since the difference between the solubility products of silver(I) bromide and silver(I) chloride is not as great as that between silver(I) iodide and silver(I) bromide, the bromide ion cannot be titrated as accurately in the presence of chloride ion as iodide in the presence of bromide. Accuracy is further lost due to the tendency of the silver(I) halides to form mixed crystals or solid solutions. This leads to a co-precipitation of a more soluble silver(I) halide (e.g., AgCl) before the precipitation of a less soluble halide (e.g., AgBr) is over. Thus the volumes of silver(I) ions solution required to titrate iodide and bromide ions are much large than the prediction based on solubility products. The errors can be minimised to 1-2% if a mixture containing equal volumes of all three halides is taken.

10.10 Concentration Cells

It is possible to construct a galvanic cell from two half-cells, both of which have copper electrodes dipping in copper(II) ion solution, provided each half-cell has a different concentration of copper(II) ions. This is because the electrode potential depends on the copper(II) ion concentration, as shown in Table 10.2. Such a cell is known as a *concentration cell.* The half-cell with the lower activity of copper(II) ions will act as the anode. The cell reaction takes place in order to reduce the difference in concentrations. When the two concentrations are equal, equilibrium is achieved. If the two half-cells are linked by a salt-bridge, the cell is known as a *concentration cell without transference.* This cell can be written as

$$Cu(s) \mid Cu^{2+}(aq), a_1 \parallel Cu^{2+}(aq), a_2 \mid Cu(s)$$

where a_1 and a_2 denote the different activities of copper(II) ions in the two half-cells.

The Nernst equation for this cell can be written as

$$E_{cell} = E_{cathode} - E_{anode}$$

$$= \{ E^0_{Cu2+/Cu} + \frac{RT}{2F} \ln a_2 \} - \{ E^0_{Cu2+/Cu} + \frac{RT}{2F} \ln a_1 \}$$

$$= \frac{RT}{2F}(\ln a_2 - \ln a_1)$$

$$E_{cell} = \frac{RT}{2F} \ln \frac{a_2}{a_1} \quad (10.37)$$

If the two half-cells of a concentration cell are separated by an open boundary, i.e., there is no salt bridge linking them then the cell is known as a *concentration cell with transference.* A potential difference develops across the boundary between the two solutions. This potential difference is called the *liquid junction potential, Ej.* To understand why this potential develops, consider two silver-silver chloride electrodes, one dipped in concentrated hydrochloric acid solution of activity a_1, the other dipped in dilute hydrochloric acid solution of activity a_2 (Figure 10.11). The hydrogen ions and chloride ions in the more concentrated solution diffuse into the more dilute solution. Since the mobility of the H^+ ion is much more than that of the Cl^- ion, it

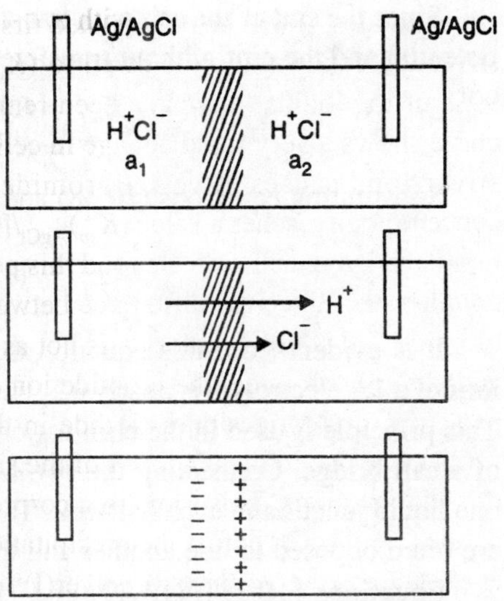

Figure 10.11: The liquid junction in a concentration cell.

diffuses faster. The fast moving H^+ ions lead ahead of the slower Cl^- ions. This results in a separation of charges and an electrical double layer at the boundary of the two solutions. The potential difference due to the electrical double layer slows down the faster moving H^+ ions and speeds up the slower Cl^- ions. After some time, equilibrium is achieved and the two ions diffuse at the same speed.

The additional potential difference created due to the liquid junction shows up in the measurement of the cell potential. Choosing the half-cell with activity a_1 as the anode, the cell can be expressed in terms of cell notation as

$$\text{Ag(s)} \mid \text{AgCl(s)} \mid \text{Cl}^-\text{(aq)}, a_1 \text{ M } \vdots\ \text{Cl}^-\text{(aq)}, a_2 \mid \text{AgCl(s)} \mid \text{Ag(s)}$$

where the dotted vertical line, $\vdots$, represents the liquid junction. It is possible to derive an expression for the emf of this cell and is available in any standard book on Physical Chemistry. The emf of this cell is given by

$$E_{wt} = -2t_+ \frac{RT}{2F} \ln \frac{a_2}{a_1} \tag{10.38}$$

where the subscript 'wt' denotes 'with transference'

t_+ is the transport number of the H^+ ion.

For a detailed discussion on transport numbers, see Section 9.4.1 of this book.

If the liquid junction did not exist, the emf of the cell would be

$$E_{wot} = -\frac{RT}{F} \ln \frac{a_2}{a_1} \tag{10.39}$$

where the subscript 'wot' denotes 'without transference'.

Since the emf of the cell with transference is the sum of the liquid junction potential and the emf without transference,

$$E_{wt} = E_{wot} + E_j$$

$$E_j = E_{wt} - E_{wot} \tag{10.40}$$

Substituting equations (10.38) and (10.39) into (10.40), we get

$$E_j = (1 - 2t_+) \frac{RT}{F} \ln \frac{a_2}{a_1} \tag{10.41}$$

It is evident from this equation that when the transport numbers of the ions of a 1,1 electrolyte is nearly 0.5, the liquid junction potential is negligible. This principle is used in the elimination of liquid junction potential by means of a salt bridge. Connecting the two half-cells with a salt bridge introduces one liquid junction in each half-cell. The potentials of the two liquid junctions are often opposed to one another and the net junction potential is very small.

Equations (10.38), (10.39) and (10.41) are derived with respect to a concentration cell composed of electrodes that are reversible to anions, and for a 1,1 electrolyte. Table 10.7 summarises the equations generally applicable to concentration cells categorised according to the type of ion to which their electrodes are reversible as well as the type of electrolyte.

Table 10.7: Emf of concentration cells with and without transference

Electrodes reversible to	*Emf of cell / (V)*	
	with transference	*without transference*
Anions	$-t_+ \frac{(\nu_+ + \nu_-)}{\nu_- z_-} \frac{RT}{F} \ln \frac{a_2}{a_1}$	$-\frac{RT}{z_- F} \ln \frac{a_2}{a_1}$
Cations	$-t_+ \frac{(\nu_+ + \nu_-)}{\nu_+ z_+} \frac{RT}{F} \ln \frac{a_2}{a_1}$	$+\frac{RT}{z_+ F} \ln \frac{a_2}{a_1}$

ν_+ and ν denote respectively the number of cations and anions furnished on dissociation of the electrolyte. z_+ and z_- denote respectively the oxidation number of the cations and anions. a_1 and a_2 are the mean ionic activity coefficients of the ions in the anode and cathode compartments respectively.

10.11 Answer to Puzzle

A puzzle was given at the beginning of this chapter. The colourless solution is a solution containing zinc ions. They oxidise copper metal atoms on the surface of the copper rod to copper(II) ions which enter the solution.

$$Zn^{2+} + Cu \rightarrow Zn + Cu^{2+}$$

The zinc metal atoms formed deposit on the surface of the copper rod and give it a silvery colour. When this rod is heated strongly, zinc and copper alloy to form brass, which is golden-yellow in colour.

EXPERIMENTS

Experiment 10.1

Aim: To determine the solubility product of silver chloride using a chemical cell.

Requirements

0.01 M $AgNO_3$, 0.02 M KCl, silver electrode, saturated calomel electrode, ammonium nitrate salt bridge, potentiometer assembly.

Theory

The chemical cell is represented as

$$^{-}\ \text{SCE} \parallel AgCl_{(s)},\ KCl_{(aq)}|Ag_{(s)}\ ^{+}$$

The emf of this cell is given by

$$E_{cell} = E^0_{\text{oxi (SCE)}} + E^0_{\text{red (Ag+/Ag)}} + 2.303RT/F \log(a_{Ag+})$$
$$= -0.244 + 0.799 + 0.05916 \log (a_{Ag+})$$

$$= 0.555 + 0.05916 \log (a_{Ag+})$$

The emf of the cell is measured by varying the concentration of KCl in the right-hand (cathode) compartment. Thus the activity of Ag^+ ions in the cathode compartment can be calculated for each value of KCl concentration by using the Nernst equation. The solubility product of the sparingly soluble silver chloride can be obtained by

$$K_{sp} = a_{Ag+} \cdot a_{Cl^-}$$

Procedure

Prepare 0.01M, 0.005M, 0.0025M and 0.001 M solutions of KCl from the 0.02M KCl solution.

Set-up and standardise the potentiometer. To set-up the cell take 50 cm^3 of 0.001 M KCl solution into a beaker and add one drop of $AgNO_3$ solution to the KCl solution in order to saturate the latter with AgCl. Immerse the clean silver electrode in the solution. In another beaker, dip the calomel electrode in saturated KCl solution. Connect the electrodes to the potentiometer and determine the emf.

Vary the solution in the cathode (right hand) compartment by replacing it successively with 0.0025M, 0.005M, 0.01M and 0.02M KCl solutions and record the corresponding emf.

Calculate the activity of Ag^+ ions in the anode compartment for each of the measured cell potentials using the Nernst equation. Calculate the solubility product of AgCl. The mean ionic activity coefficients of KCl for various concentrations of KCl are given below:

Concentration of KCl	0.02	0.01	0.005	0.0025	0.001
Activity coefficient at 298K	00.869	0.899	0.925	0.945	0.964

Observations and calculations

Cell potential	Cathode Compartment	
E_{cell} (V)	Conc. of KCl (M)	Vol of KCl (cm^3)
	0.02	50
	0.01	50
	0.005	50
	0.0025	50
	0.001	50

According to the Nernst equation,

$$E_{cell} = 0.555 + 0.05916 \log (a_{Ag+})$$

Therefore,

$$a_{Ag+} = \text{antilog} [(E_{cell} - 0.555)/0.05916]$$

Conc. of KCl m_{KCl} (M)	$(\gamma_{\pm})_{KCl}$	Cell emf E_{cell} (V)	$\frac{(E_{cell} - 0.555)}{0.05916}$	a_{Ag+}	$a_{Cl-} = m_{KCl}(\gamma_{\pm})_{KCl}$	$K_{sp} = a_{Ag+} . a_{Cl-}$
0.02	0.869					
0.01	0.899					
0.005	0.925					
0.0025	0.945					
0.001	0.964					

Mean K_{sp} = ________

Result

The solubility product of silver chloride is ________.

Experiment 10.2

Aim: Determination of pK values of maleic acid/malonic acid by potentiometric titration with sodium hydroxide using glass electrode.

Theory

Maleic acid and malonic acids are both dibasic acids. They dissociate in two steps and thus have two dissociation constants. The first dissociation is similar to that of a strong acid and the second one similar to that of a weak acid. At the start of the pH metric titration, the pH is in the acidic range and increases slowly with successive additions of NaOH. Near the completion of the first dissociation, there is a sharp rise in the pH of the solution. After the first end point, the pH again rises slowly till the second dissociation when there is once again a sharp rise in the pH of the solution. From the two inflexion points in the titration curve, the pK values of the acid can be determined.

Requirements

pH meter, buffer solutions of pH 4.01 and 9.18, 0.1N NaOH, 0.1N maleic/malonic acid solution.

Procedure

Set-up and standardise the pH meter using the given buffer solutions. Pipette out 10 cm^3 of maleic/malonic acid solution in a beaker. Immerse the electrode in this solution and determine its pH. Titrate against previously standardised sodium hydroxide from the burette, adding 0.5 cm^3 at a time. The volume of sodium hydroxide added should be reduced to 0.1 cm^3 towards the end points. Take five or six more readings after the second neutralisation point.

Plot a graph of

(i) pH versus volume V of the titrant

(ii) ΔpH/ΔV versus mean V

Determine the pK values and the dissociation constants K of the acid. Also determine the amount of maleic/malonic acid present in the given solution.

Observations and calculations

Solution in burette: 0.1N NaOH

Solution in beaker: 10 cm^3 of 0.1N maleic/malonic acid

Vol of NaOH V (cm^3)	pH	ΔpH/ΔV	Mean V (cm^3)

From the graph of pH versus volume V of NaOH added, the two end points are:

V_1 = ____cm^3, V_2 = ____cm^3,

$$pK_1 = \text{pH at } (V_1/2)$$

$$pK_2 = \text{pH at } ((V_2 - V_1)/2 + V_1)$$

Volume of NaOH required to completely neutralise the acid present = V_2 cm^3.

Calculations for malonic acid

Molar mass of malonic acid (HOOC – CH_2 – COOH) = 104.06 g mol^{-1}

1000 cm^3 of 1N NaOH ≡ 52.03 g of malonic acid

Therefore, V_3 cm^3 of 0.1N NaOH = 52.03 V_3 × 0.1/1000 g of malonic acid

= A g of malonic acid

Now A g of malonic acid is present in 10 cm^3 of given solution.

Therefore, amount of malonic acid present in 100 cm^3 of the given solution = 10A g

Calculations for maleic acid

Molar mass of maleic acid (HOOC – CH=CH – COOH) = 110.08 g mol^{-1}

1000 cm^3 of 1N NaOH = 55.04 g of maleic acid

Therefore, V_3 cm^3 of 0.1N NaOH = 55.04 V_3 × 0.1/1000 g of maleic acid

= B g of maleic acid

Now B g of malonic acid is present in 10 cm^3 of given solution.

Therefore, amount of malonic acid present in 100 cm^3 of the given solution = 10B g

Result

First dissociation constant pK_1 = _______

Second dissociation constant pK_2 = _______

Amount of maleic/malonic acid present in the given solution = ______ g

Experiment 10.3

Aim: Determination of pK values of phosphoric acid by potentiometric titration with sodium hydroxide using glass electrode.

Theory

When a weak monobasic acid is titrated with a strong base such as sodium hydroxide, the pH of the solution at any stage of neutralisation prior to the end point is given by Henderson's equation:

$$\text{pH} = pK_a + \log (C_{salt}/C_{acid}) + \log (\gamma_{\pm\,salt}/\gamma_{\pm acid})$$

where C_{salt} is the concentration of the salt formed

C_{acid} is the concentration of unneutralised acid

$\gamma_{\pm\,salt}$ and $\gamma_{\pm acid}$ are the activity coefficients of the respective species.

The pK_a value of a monobasic acid is approximately equal to the pH value at the point corresponding to the addition of one half of the base required to reach equivalence point i.e. when $C_{salt}/C_{acid} = 1$.

For a dibasic acid, if the second pK_a value is approximately hundred times greater than the first then the second break in the curve can be easily detected. However, if the ratio K_1/K_2 is less than 100 then it is difficult and often impossible to detect a break in the curve at equivalence point.

The reasoning can be extended to a triprotic acid. In case of phosphoric acid, the third stage of neutralisation is very weak. The solution is too alkaline for sodium hydroxide to produce any significant difference of pH and hence the curve is very flat. To get around this difficulty, calcium chloride is added to the solution at some point after the second equivalence point. Calcium chloride reacts with disodium hydrogen phosphate present to precipitate calcium phosphate and a quantity of hydrochloric acid exactly equivalent to the acid content of disodium hydrogen phosphate.

$$3CaCl_2 + 2Na_2HPO_4 \rightarrow 4NaCl + Ca_3(PO_4)_2\downarrow + 2HCl$$

Due to the appearance of the strong acid the pH of the solution falls, thus making possible further titration until a third break corresponding to complete neutralisation results.

Requirements

0.5N NaOH, 100 cm^3 of 0.3N H_3PO_4, pH meter, buffer solutions of pH 4 and 9, solid calcium chloride dihydrate, $CaCl_2.2H_2O$.

Procedure

Set-up and standardise the pH meter using the given buffer solutions. Pipette out 50 cm^3 of phosphoric acid solution in a beaker. Immerse the electrode in this solution and determine its pH. Titrate against previously standardised sodium hydroxide from the burette, adding 0.5 cm^3 at a time. The volume of sodium hydroxide added should be reduced to 0.1 cm^3 towards the end points. Add calcium chloride after the second neutralisation point till there is complete precipitation of calcium phosphate. Add sodium hydroxide and find the third neutralisation point. Take five or six more readings after this neutralisation point.

Plot a graph of

(i) pH versus volume V of the titrant

(ii) $\Delta pH/\Delta V$ versus mean V

Determine the pK values and the dissociation constants K of the acid. Also determine the amount of phosphoric acid present in the given solution.

Observations and calculations

Solution in burette: 0.5N NaOH

Solution in beaker: 50 cm^3 of 0.3N H_3PO_4

Vol. of NaOH V (cm^3)	pH	$\Delta pH/\Delta V$	Mean V (cm^3)

From the graph of pH versus volume V of NaOH added, the three end points are:

$V_1 =$ ____cm^3, $V_2 =$ ____cm^3, $V_3 =$ ____cm^3,

$pK_1 = \text{pH at } (V_1/2)$

$pK_2 = \text{pH at } ((V_2 - V_1)/2 + V_1)$

$pK_3 = 2\,(\text{pH at } V_2) - pK_2$

Volume of NaOH required to completely neutralise the acid present = V_3 cm^3

Now 1000 cm^3 of 1N NaOH = 32.67 g of H_3PO_4

Therefore, V_3 cm^3 of 0.5N NaOH = 32.67 $V_3 \times 0.5/1000$ g of H_3PO_4

= A g of H_3PO_4

Now A g of H_3PO_4 is present in 50 cm^3 of given solution.

Therefore, amount of H_3PO_4 present in 100 cm^3 of the given solution = 2A g

Result

First dissociation constant pK_1 = _______

Second dissociation constant pK_2 = _______

Third dissociation constant pK_3 = _______

Amount of phosphoric acid present in the given solution = ______ g

Experiment 10.4

Aim: Determination of the concentrations/amounts of iodide, bromide and chloride in the mixture by potentiometric titration with silver nitrate.

Theory

The most important precipitation titrations are those involving salts of silver, e.g., silver halides. The titration can be carried out potentiometrically using a piece of pure silver wire as indicator electrode.

The electrode potential of the silver electrode in equilibrium with silver ions is by definition the potential of the cell in which the electrode on the left is the standard hydrogen electrode and that on the right is the Ag/Ag^+ electrode. The cell is represented as:

$$^{-}\ Pt_{(s)} \mid H_{2(g)} \mid H^{+}_{(aq)} \parallel Ag^{+}_{(aq)} \mid Ag_{(s)}\ ^{+}$$
$$a = 1$$

The cell reaction is $H_{2(g)} + 2Ag^{+}_{(aq)} \rightarrow 2H^{+}_{(aq)} + 2Ag_{(s)}$

The half reaction at the right hand electrode is

$$Ag^{+}_{(aq)} + e^{-} \rightarrow Ag_{(s)}$$

The zero current cell potential is given by the Nernst equation.

$$E = E^{\circ}_{Ag+/Ag} - RT/nF \ln Q$$

where $$Q = a^{2}_{H+}/F_{(H_2/P^{\circ})}a^{2}_{(Ag+)}$$

When the hydrogen electrode has its standard composition, $a_{H+} = 1$ and $F_{H2} = 1$ P but the Ag^+/Ag electrode has an arbitrary composition under these conditions.

Therefore, $$Q = 1/a^{2}_{Ag+}$$

And the cell potential can be regarded as arising solely from the Ag^+/Ag electrode. Thus

$$E_{Ag+/Ag} = E^{\circ}_{Ag+/Ag} - RT/2F \ln(1/a^{2}_{Ag+})$$
$$= E^{\circ}_{Ag+/Ag} - RT/F \ln(1/a_{Ag+})$$

The experimental cell comprises of a half-cell with the silver electrode dipped in the solution being analysed. The other half-cell consists of the standard calomel electrode immersed in saturated potassium nitrate solution. The two half cells are connected by means of an agar bridge containing 3% potassium nitrate solution.

The cell is represented as

$$^{-}Hg_{(s)} \mid Hg_2Cl_{2(s)} \mid KCl_{(aq)} \parallel \text{halide soln} \mid Ag_{(s)}\ ^{+}$$

The potential of this cell at 298 K is given by

$$E = E^{\circ}_{Ag+/Ag} - E_{SCE}$$
$$= E^{\circ}_{Ag+/Ag} - E_{SCE} + 0.0592 \log a_{Ag+}$$

When silver nitrate solution is added to halide solution during early stages of titration, only silver iodide precipitates. Under equilibrium conditions, the precipitation of silver bromide will not begin until all but 0.02% of the iodide originally present has been precipitated. Then precipitation of silver bromide occurs and again no silver chloride until all but 0.3% of bromide ions has been removed from the solution. The cell potential depends on log a_{Ag+} or, less accurately, on log $[Ag^+]$ and hence it gradually increases during the precipitation.

The ion concentration at the equivalence point is determined by the solubility product of the sparingly soluble halide formed during titration. There is rapid change in the concentration of silver ions and hence in overall cell potential near the equivalence point.

Requirements

Potentiometer, silver electrode, saturated calomel electrode, salt bridge (NH_4NO_3), 0.1M silver nitrate, 0.1N KCl solution, 2M HNO_3, 100 cm^3 of halide solution.

Procedure

Set-up and standardise the potentiometer. Set-up the cell and take 25 cm^3 of halide solution into a beaker and add 2-3 cm^3 of 2M HNO_3 solution. Immerse the clean silver electrode in the solution. Dip the calomel electrode in saturated KCl solution. Connect the electrodes to the potentiometer and determine the emf. Record the polarity of the electrodes. Run in the silver nitrate solution from the burette. Stir well and determine the emf after each addition. The volume added should be decreased to 0.1cm^3 towards each end point.

Plot the following graphs:

(i) Emf E versus volume V

(ii) $\Delta E/\Delta V$ versus mean volume V'

(iii) $\Delta^2 E/\Delta V^2$ versus mean volume V''

Determine the concentration and weight of each halide in the given solution.

Observations and calculations

Solution in burette: 0.1N $AgNO_3$

Solution in titration vessel: 25 cm^3 of halide mixture

Vol. of $AgNO_3$ V (cm^3)	Cell emf E (mV)	$\Delta E/\Delta V$	mean volume V'	$\Delta^2 E/\Delta V^2$	mean volume V''
0.0					
0.2					
..					

From the graph of $\Delta^2E/\Delta V^2$ versus mean volume V'' of $AgNO_3$ added, the three end points are:

$V_1 =$ ______ cm^3, $V_2 =$ ______+ cm^3, $V_3 =$ ______ cm^3

Normality of halide mixture

= Normality of $AgNO_3 \times V_3$/Volume of halide mixture taken

$= 0.1V_3/25 = 0.004V_3\ N = B\ N$

Volume of $AgNO_3$ required for complete precipitation of AgI $= V_1\ cm^3$

Therefore, normality of I^- ions $= B\ V_1/V_3 = C_1\ N$

Now 1000 cm^3 of 1 $N\ I^-$ = 1 gram equivalent of I^- = 126.90 g

Therefore, $V_1\ cm^3$ of $C_1\ N\ I^- = 126.90\ V_1\ C_1/1000$ g of I^-

Volume of $AgNO_3$ required for complete precipitation of AgBr

$= V_2 - V_1\ cm^3$

Therefore, normality of Br^- ions $= B\ (V_2 - V_1)/V_3 = C_2 N$

Now 1000 cm^3 of 1 N $Br^- \equiv$ 1 gram equivalent of Br^- = 79.90 g

Therefore, $(V_2 - V_1)\ cm^3$ of C_2 N $Br^- = 79.90(V_2 - V_1)C_2/1000$ g of Br^-

Volume of $AgNO_3$ required for complete precipitation of AgCl

$= V_3 - V_2\ cm^3$

Therefore, normality of Cl^- ions $= B\ (V_3 - V_2)/V_3 = C_3\ N$

Now 1000 cm^3 of 1 N $Cl^- \equiv$ 1 gram equivalent of Cl^- = 35.453 g

Therefore, $(V_3 - V_2)\ cm^3$ of C_3 N $Cl^- = 35.453(V_3 - V_2)C_3/1000$ g of Cl^-

Result

Halide	Concentration (M)	Amount in the given soln (mol)	Weight in the given soln (g)
Chloride			
Bromide			
Iodide			

Aim: Determination of the solubility product of silver chloride potentiometrically using a concentration cell.

Requirements

Potentiometer assembly, two silver electrodes, KNO_3 or NH_4NO_3 salt bridge, 0.1M KCl, 0.01M $AgNO_3$

Procedure

Set-up the potentiometer assembly and standardise it. Clean the silver electrodes and immerse them both in 0.01M $AgNO_3$ solution. Connect the two half-cells through a salt bridge and measure the potential of the cell. The cell potential should be zero.

Prepare 0.0005M, 0.001M and 0.005M solutions of silver nitrate from the 0.01M $AgNO_3$ solution.

Dip one electrode in 0.01M $AgNO_3$ solution contained in a beaker and the other in 0.01M KCl solution contained in a beaker. Add a fraction of a drop of $AgNO_3$ solution to the KCl solution in order to saturate the latter with AgCl. Vary the solution in the cathode (right hand) compartment by replacing it successively with 0.0005M, 0.001M, 0.0025M and 0.005M $AgNO_3$ solutions and record the corresponding emf.

Calculate the activity of Ag^+ ions in the anode compartment for each of the measured cell potentials using the Nernst equation. Taking the activity coefficient of Cl^- ions in 0.01M KCl as 0.902, calculate the solubility product of AgCl. The activity coefficients for various concentrations of Ag^+ ions are given below:

Conc. of $AgNO_3$	0.01	0.005	0.0025	0.001
Activity coefficient at 298K	0.892	0.924	0.945	0.964

Observations and calculations

Cell potential E(V)	Anode Compartment	Cathode Compartment	
	Vol. of 0.01M KCl saturated with AgCl (cm^3)	Conc. of $AgNO_3$	Vol of $AgNO_3$ (cm^3)
	50	0.01	50
	50	0.005	50
	50	0.0025	50
	50	0.001	50

According to the Nernst equation,

$$E = 2.303RT/F \ \log[(a_{Ag+})_1 / (a_{Ag+})_2]$$

where $(a_{Ag+})_1$ and $(a_{Ag+})_2$ are the activities of Ag^+ ions in the anode and cathode compartments respectively.

$$(a_{Ag+})_1 = (a_{Ag+})_2 /[\text{antilog}(E/0.05916)]$$
$$= m_2(\gamma_\pm)_2/[\text{antilog}(E/0.05916)]$$

Conc. of $AgNO_3$ m_2 (M)	$(\gamma_\pm)_2$	Cell emf E (V)	Antilog (E/0.05916)	$(a_{Ag+})_1$ $\times 10^{-8}$	$a_{Cl-} = \gamma_{Cl-} C_{Cl-}$	$K_{sp} = a_{Ag+} . a_{Cl-}$ $\times 10^{-10}$
0.001						
0.0025						
0.005						
0.01						

Mean K_{sp} = _____

Solubility = $\sqrt{K_{sp}}$ = _____ mol dm^{-3}

Result

The solubility product of silver chloride is ______

Solubility of silver chloride = ______

(Solubility of silver chloride at 25 °C= 1.76×10^{-5} mol dm^{-3})

Experiment 10.5

Aim: To find the empirical formula of silver-ammonia complex by potentiometric method.

Requirements

0.01N $AgNO_3$, 1N NH_3, potentiometer, silver electrodes, agar bridge.

Procedure

Set-up the potentiometer circuit and calibrate. Clean the two silver electrodes and check that when they are immersed in the same silver nitrate solution, no emf is observed. Set-up the concentration cell shown, taking the usual precautions of washing and measure the emf in each case.

Observations and calculations

Set I

Emf *E* (V)	Anode compartment			Cathode compartment		
	Vol. of 0.01M $AgNO_3$ (cm^3)	Vol. of 1M NH_3 (cm^3)	Vol. of H_2O (cm^3)	Vol. of 0.01M $AgNO_3$ (cm^3)	Vol. of 1M NH_3 (cm^3)	Vol. of H_2O (cm^3)

Emf *E* (V)	$(C_{Ag+})_1$ (M)	$(C_{Ag+})_2$ (M)	$x = (0.05916/E)\log [(C_{Ag+})_2/(C_{Ag+})_1]$

Mean x = ______

Set II

Emf *E* (V)	Anode compartment			Cathode compartment		
	Vol. of 0.01M $AgNO_3$ (cm^3)	Vol. of 1M NH_3 (cm^3)	Vol. of H_2O (cm^3)	Vol. of 0.01M $AgNO_3$ (cm^3)	Vol. of 1M NH_3 (cm^3)	Vol. of H_2O (cm^3)

Emf E (V)	$(C_{NH3})_1$ (M)	$(C_{NH3})_2$ (M)	$y = xE / \{0.05916 \log [(C_{NH3})_1 / (C_{NH3})_2]\}$

Mean y = ______

Empirical formula of silver-ammonia complex is $[Ag_x(NH_3)_y]$.

Result

The empirical formula of the silver-ammonia complex by potentiometric method is __________.

Experiment 10.6

Aim: *To find the equilibrium constant of the reaction*

$$[Ag(NH_3)_2]^+ \;\square\; Ag^+ + 2NH_3$$

Requirements

0.1M $AgNO_3$, 2M NH_3, potentiometer, silver electrodes, agar bridge, saturated KNO_3 solution.

Procedure

Set-up and standardise the potentiometer circuit using Weston standard cadmium cell. Check the two silver electrodes by dipping in the same silver nitrate solution and measuring the emf. The emf must be zero.

Prepare 0.1M $AgNO_3$ solution in 2M NH_3 by mixing equal volumes of 0.1M $AgNO_3$ and 2M NH_3 solution. Set-up the concentration cell shown, taking the usual precautions of washing and measure the emf in each case. Repeat the determination for different dilutions of silver nitrate.

Observations and calculations

Set I

Emf E (V)	Anode compartment			Cathode compartment	
	Conc. of $AgNO_3$ (cm^3)	Vol of $AgNO_3$ (cm^3)	Vol of 2MNH_3 (cm^3)	Vol of 0.05M $AgNO_3$ (cm^3)	Vol of H_2O (cm^3)

Concentration of Ag^+ ions in the cathode compartment = 0.05 × 10/50 = 0.01 = m_2

$\gamma_2 = 0.898$

Emf E (V)	$(a_{Ag+})_1 = m_2\gamma_2/\{\text{antilog}\ (E/0.05916)\}$	Conc. of Ag^+ originally present x (M)	Conc. of NH_3 originally present γ (M)	Conc. of free NH_3 $(y - 2x) = c$ (M)

$\gamma\ [Ag(NH_3)_2]^+ = \text{antilog}\ (-0.509\sqrt{x})$	$K = (a_{Ag+})_1 (C_{NH3})^2\ / x . \gamma\ [Ag(NH_3)_2]^+$

Mean K = _____

Result

The equilibrium constant of the reaction $[Ag(NH_3)_2]^+$ = _______

Experiment 10.7

Aim: To determine the acidic and basic dissociation constants of an amino acid and hence its isoelectric point.

Requirements

pH meter, 0.1 M NaOH, 0.2 M HCl, 0.1 N glycine

Theory

Amino acids are ampholytes, i.e., they have separate acidic and basic groups. They can be generally represented as $H_2N - R - COOH$. Addition of acid results in neutralisation of $-NH_2$ group.

$$H_2N - R - COOH + H_3O^+ \rightarrow H_3N^+ - R - COOH + H_2O$$

Addition of a strong base gives

$$H_2N - R - COOH + OH^- \rightarrow H_2N - R - COO^- + H_2O$$

These can be determined by a simplified technique where an amino acid is considered as a dibasic acid. Consider the hydrochloride of glycine:

$$Cl\text{–}H_3N^+ - R - COOH$$

When this is neutralised by alkali there are two stages of reaction corresponding to the two stages of neutralisation of dibasic acid:

$$H_3N^+ - R - COOH + OH^- \rightarrow H_3N^+ - R - COO^- + H_2O$$
$$H_3N^+ - R - COO^- + OH^- \rightarrow H_2N - R - COO^- + H_2O$$

Replacing $H_3N^+ - R - COO^-$ by RH^+, $H_3N^+ - R - COOH$ by RH_2^+ and $H_2N - R - COO^-$ by R^-, we get

$$RH_2^+ + H_2O \longleftrightarrow H_3O^+ + RH^+$$
$$RH^+ + H_2O \longleftrightarrow H_3O^+ + R^-$$

The dissociation constants of the two stages are

$$K_1 = \frac{a_{H+}\, a_{RH+}}{a_{RH_2+}}$$

and

$$K_2 = \frac{a_{H+}\, a_{R}}{a_{RH+}}$$

$$pK_1 = pH - \log \frac{C_{RH+}}{C_{RH2+}} - \log \frac{\gamma_{RH^+}}{\gamma_{RH_2^+}}$$

$$pK_2 = pH - \log \frac{C_{R}}{C_{RH+}} - \log \frac{\gamma_{R}}{\gamma_{RH+}}$$

If a solution is made up of *c* equivalents of neutral glycine and *a* equivalents of strong acid then

$$C_{RH+} = c - a + C_{H+}$$

$$C_{RH2+} = a - C_{H+}$$

$$pK_1 = pH - \log \frac{(C-1)}{a - C_{H+}} - \log \frac{\gamma_{RH+}}{\gamma_{RH2+}}$$

The last term in this equation can be taken as zero.

If a solution is made up of *c* equivalents of neutral glycine and *b* equivalents of a strong base then

$$C_{RH+} = c - b + C_{OH-}$$

$$C_{R-} = b - C_{OH-}$$

$$pK_2 = pH - \log \frac{(C-1)}{b - C_{OH-}} - \log \frac{\gamma_{R-}}{\gamma_{RH+}}$$

The last term in this equation can be taken as zero.

In order to determine pK_1 and pK_2 a solution is made up of known amounts of neutral amino acid and either strong base or strong acid. The pH of this solution is determined from which values of C_{H+} and C_{OH-} can be calculated.

Procedure

Pipette out 10 cm^3 of 0.1 N glycine solution in a 100 cm^3 beaker. Dip the glass electrode of the pH meter in the solution. Fill the burette with previously standardised 0.1 M NaOH solution. Titrate the amino acid solution with NaOH. Note down the pH value of the solution after each addition of 1 cm^3 of standard NaOH solution from the burette. Take 10 such readings.

Pipette out another 10 cm^3 of 0.1 N glycine solution in a 100 cm^3 beaker. Dip the glass electrode of the pH meter in the solution. Fill the burette with previously standardised 0.2 M HCl solution. Titrate the amino acid solution with HCl. Note down the pH value of the solution after each addition of 1 cm^3 of standard HCl solution from the burette. Take 10 such readings.

Calculate the pK_1 and pK_2 values of the amino acid and hence the isoelectric point of the amino acid.

Observations and calculations

I. Determination of pK$_1$

Vol. of HCl added V_{HCl} (cm^3)	pH	Conc. of amino acid C (M)	$a = \frac{V_{HCl} N_{HCl}}{V_{TOTAL}}$	[H+] (M)	$x = \left(\frac{C}{a - [H^+]}\right) - 1$	log x	pK_1 = pH – log x
1.0							
2.0							
…							
10.0							

Mean pK_1 = ____

II. Determination of pK$_2$

Vol. of NaOH added V_{NaOH} (cm^3)	pH	Conc. of amino acid C (M)	$b = \frac{V_{NaOH} N_{NaOH}}{V_{TOTAL}}$	$[OH^-]$ = antilog (pH – 14) (M)	$x = \left(\frac{C}{b - [OH^-]}\right) - 1$	log x	pK_2 = pH + log x
1.0							
2.0							
…							
10.0							

Mean pK_2 = ____

Isoelectric point $pI = \frac{pK_1 + pK_2}{2}$ = ________

Results

1. pK_1 of glycine = ______
2. pK_2 of glycine = ______
3. The isoelectric point pI of glycine = _____.

(Expected value: $pK_1 = 2.34$ $pK_2 = 9.60$ pI = 5.97)

Experiment 10.8

Aim: Determination of Hammett constant of o–, m–, p–amino/nitro benzoic acid by pH measurements.

Requirements

50% ethanol-water mixture, 0.02M benzoic acid in 50% ethanol-water mixture, 0.02M *p*-aminobenzoic acid in 50% ethanol-water mixture, 0.02M *p*-nitrobenzoic acid in 50% ethanol-water mixture, 0.2N NaOH, pH meter.

Theory

The effect of substituents at meta and para position of aromatic acids on its equilibrium constant can be studied by using the Hammett equation. The equation is applicable to aromatic compounds having the same reaction centre present at a side-chain and having a substituent in the meta or para position to this reaction centre. According to this equation, the equilibrium constant of the substituted compound is related to the equilibrium constant of the unsubstituted compound along with two constants σ and ρ. It can be given as

$$\log K = \log K^0 + \sigma\rho$$
$$-\log K = -\log K^0 - \sigma\rho$$
$$pK = pK^0 - \sigma\rho$$
$$pK^0 - pK = \sigma\rho$$

σ is known as substituent constant because it depends only on its substituent. ρ is known as the reaction constant, which mainly depends on the solvent used. The reactivity of the unsubstituted compound changes by substituent, which changes the availability of electrons at the seat of reaction.

Therefore by knowing the values of σ for different substituents, the change in acidity of the compound can be shown. Substituents with positive σ-values are stronger electron attractors and those with negative σ-values are weak electron attractors. Therefore the reactions with positive σ-values are accelerated by electron withdrawal from benzene ring, whereas those with negative σ-values are retarded by electron withdrawal.

Procedure

Prepare 50% ethanol-water mixture. Prepare 0.02M solutions of benzoic acid, *p*-aminobenzoic acid and *p*-nitrobenzoic acid in the 50% ethanol-water mixture. Pipette out 50 cm^3 of 0.02M benzoic acid solution in a 150 ml beaker and titrate against exact 0.2N NaOH solution. Note down the pH values for addition of every 0.5 cm^3 of NaOH solution. Repeat the procedure for *p*-aminobenzoic acid and *p*-nitrobenzoic acid.

Plot a graph of pH against volume of NaOH added for all three acids. Determine the dissociation constant of each acid from its respective graph. The *pK* value will be the value of pH at half-neutralisation point. Calculate the Hammett constant by using the equation

$$pK^0 - pK = \sigma\rho$$

where pK^0 is the dissociation constant of benzoic acid

pK is the dissociation constant of the substituted benzoic acid

ρ is 1.65 for 50% ethanol-water mixture

Observations

Solution in burette: 0.2N (exact) NaOH

Solution in beaker: 50 cm^3 0.02N benzoic acid

$V(cm^3)$	pH
0.0	
0.5	.
...	...

Solution in burette: 0.2N (exact) NaOH

Solution in beaker: 50 cm^3 0.02N 4-nitrobenzoic acid

$V(cm^3)$	pH
0.0	

Solution in burette: 0.2N (exact) NaOH

Solution in beaker: 50 cm^3 0.02N 4-aminobenzoic acid

$V(cm^3)$	pH
0.0	

Calculations

$$\begin{aligned} \text{NaOH} &\equiv \text{Benzoic acid} \\ N_1V_1 &= N_2V_2 \\ 0.2 \times V_1 &= 0.02 \times 50 \\ V_1 &= 5\ cm^3 \end{aligned}$$

Volume of NaOH required to neutralise benzoic acid = 5 cm^3

From the graph of pH against volume of NaOH added for benzoic acid,

pK of benzoic acid = pH at $V_1/2$ = pH at 2.5 cm^3 = _____

Thus pK^0 = ____

$$\begin{aligned} \text{NaOH} &\equiv p\text{-nitrobenzoic acid} \\ N_1V_1 &= N_2V_2 \\ 0.2 \times V_1 &= 0.02 \times 50 \\ V_1 &= 5\ cm^3 \end{aligned}$$

Volume of NaOH required to neutralise *p*-nitrobenzoic acid = 5 cm^3

From the graph of pH against volume of NaOH added for *p*-nitrobenzoic acid,

pK of *p*-nitrobenzoic acid = pH at $V_1/2$ = pH at 2.5 cm^3 = _____

Hammett acidity constant for *p*-nitrobenzoic acid = $(pK^0 - pK)/\rho$

= _______

$$\begin{aligned} \text{NaOH} &= p\text{-aminobenzoic acid} \\ N_1V_1 &= N_2V_2 \\ 0.2 \times V_1 &= 0.02 \times 50 \\ V_1 &= 5 \text{ cm}^3 \end{aligned}$$

Volume of NaOH required to neutralise *p*-aminobenzoic acid = 5 cm^3

From the graph of pH against volume of NaOH added for *p*-aminobenzoic acid, pK of *p*-aminobenzoic acid = pH at $V_1/2$ = pH at 2.5 cm^3 = ____

Hammett acidity constant for *p*-nitrobenzoic acid = $(pK^0 - pK)/\rho$

= ______

Result

Hammett acidity constant of *p*-aminobenzoic acid = ____

Hammett acidity constant of *p*-nitrobenzoic acid = ____

Experiment 10.9

Aim: To determine the transport numbers of Ag^+ and NO_3^- ions in silver nitrate solution potentiometrically.

Requirements

Potentiometer assembly, silver electrodes, saturated potassium nitrate solution, agar salt bridge, 0.1 M silver nitrate, 0.01 M silver nitrate.

Procedure

The concentration cell to be used is

$$-\text{Ag (s)} \mid \underset{0.01\text{ M}}{\text{AgNO}_3\text{ (aq)}} \parallel \underset{0.01\text{ M}}{\text{AgNO}_3\text{ (aq)}} \mid \text{Ag (s)} +$$

Set-up the cell by half-filling two 100 cm^3 beakers with their respective solutions and dipping a silver electrode in each solution. This cell acts as a cell with transference when the two half-cells are joined by an agar bridge made with 0.1 M silver nitrate solution. Measure the emf E_1 of this cell. The same cell acts as a cell without transference when the two half-cells are joined by an agar bridge made with saturated potassium nitrate solution. Measure the emf E_2 of this cell.

Vary the concentration of silver nitrate solution in the anode compartment and measure E_1 and E_2 for each concentration.

Calculate the transport numbers of silver and nitrate ions from the values of E_1 and E_2.

Observations and calculations

Conc. of $AgNO_3$ / (M)		Cell emf / (V)		t_{NO3-} $= E_1/2E_2$	$t_{Ag+} =$ $1 - t_{NO3-}$
in anode compartment	in cathode compartment	with transference E_1	without transference E_2		
0.01	0.1				
0.02	0.1				
0.03	0.1				

Mean t_{NO3-} = ______

Mean t_{Ag+} = ______

Result

The transport numbers of Ag^+ and NO_3^- in silver nitrate solution are ___ and ___ respectively.

Appendix

Review Questions

1. Explain why the glass electrode cannot be used beyond pH 9.
2. What is the composition of glass used to make the glass electrode?
3. What is the pH of a solution that is 10^{-9} M in HCl?
4. What is the pH of 1 M HCl?
5. What is the difference between a concentration cell and a chemical cell?

Table 10.1: Standard reduction potentials at 298 K and 101.325 kPa listed alphabetically according to the symbols of the elements

Reduction half-reaction	E^0 / V
$Ag^+ + e^- \rightarrow Ag$	+0.7996
$AgBr + e^- \rightarrow Ag + Br^-$	+0.0713
$AgCl + e^- \rightarrow Ag + Cl^-$	+0.2223
$Ag_2CrO_4 + 2e^- \rightarrow 2Ag + CrO_4^{2-}$	+0.4470
$AgF + e^- \rightarrow Ag + F^-$	+0.779
$AgI + e^- \rightarrow Ag + I^-$	–0.1522
$Al^{3+} + 3e^- \rightarrow Al$	–1.662
$Au^+ + e^- \rightarrow Au$	+1.692
$Au^{3+} + 3e^- \rightarrow Au$	+1.498
$Ba^{2+} + 2e^- \rightarrow Ba$	–2.912
$Be^{2+} + 2e^- \rightarrow Be$	–1.847

$Bi^{3+} + 3e^- \rightarrow Bi$	+0.20
$Br_2 + 2e^- \rightarrow 2Br^-$	+1.066
$BrO^- + H_2O + 2e^- \rightarrow Br^- + 2OH^-$	+0.761
$Ca^{2+} + 2e^- \rightarrow Ca$	–2.868
$Cd^{2+} + 2e^- \rightarrow Cd$	–0.4030
$Cd^{2+} + 2e^- \rightarrow Cd(Hg)$	–0.3521
$Ce^{3+} + 3e^- \rightarrow Ce$	–2.483
$Ce^{4+} + e^- \rightarrow Ce^{3+}$	+1.61
$Cl_2 + 2e^- \rightarrow 2Cl^-$	+1.3583
$Co^{2+} + 2e^- \rightarrow Co$	–0.28
$Co^{3+} + e^- \rightarrow Co^{2+}$	+1.83
$Cr^{2+} + 2e^- \rightarrow Cr$	–0.913
$Cr_2O_7^{2-} + 14H^+ + 6e^- \rightarrow 2Cr^{3+} + 7H_2O$	+1.232
$Cr^{3+} + 3e^- \rightarrow Cr$	–0.744
$Cr^{3+} + e^- \rightarrow Cr^{2+}$	–0.407
$Cs^+ + e^- \rightarrow Cs$	–2.92
$Cu^+ + e^- \rightarrow Cu$	+0.521
$Cu^{2+} + 2e^- \rightarrow Cu$	+0.3419
$Cu^{2+} + e^- \rightarrow Cu^+$	+0.153
$F_2 + 2e^- \rightarrow 2F^-$+2.866	
$Fe^{2+} + 2e^- \rightarrow Fe$	–0.447
$Fe^{3+} + 3e^- \rightarrow Fe$	–0.037
$Fe^{3+} + e^- \rightarrow Fe^{2+}$	+0.771
$[Fe(CN)_6]^{3-} + e^- \rightarrow [Fe(CN)_6]^{4-}$	+0.358
$2H^+ + 2e^- \rightarrow H_2$	0.0000
$2H_2O + 2e^- \rightarrow H_2 + 2OH^-$	–0.8277
$H_2O_2 + 2H^+ + 2e^- \rightarrow 2H_2O$	+1.776
$Hg_2^{2+} + 2e^- \rightarrow 2Hg$	+0.7973
$Hg_2Cl_2 + 2e^- \rightarrow 2Hg + 2Cl^-$	+0.2681
$Hg^{2+} + 2e^- \rightarrow Hg$	+0.851
$2Hg^{2+} + 2e^- \rightarrow Hg_2^{2+}$	+0.920
$Hg_2SO_4 + 2e^- \rightarrow 2Hg + SO_4^{2-}$	+0.6125
$I_2 + 2e^- \rightarrow 2I^-$	+0.5355
$I_3^- + 2e^- \rightarrow 3I^-$	+0.536
$In^+ + e^- \rightarrow In$	–0.14
$In^{2+} + e^- \rightarrow In^+$	–0.40
$In^{3+} + 2e^- \rightarrow In^+$	–0.443
$In^{3+} + 3e^- \rightarrow In$	–0.3382
$K^+ + e^- \rightarrow K$	–2.931

$La^{3+} + 3e^- \rightarrow La$	–2.522
$Li^+ + e^- \rightarrow Li$	–3.0401
$Mg^{2+} + 2e^- \rightarrow Mg$	–2.372
$Mn^{2+} + 2e^- \rightarrow Mn$	–1.185
$Mn^{3+} + e^- \rightarrow Mn^{2+}$	+1.5415
$MnO_2 + 4H^+ + 2e^- \rightarrow Mn^{2+} + 2H_2O$	+1.224
$MnO_4^- + 8H^+ + 5e^- \rightarrow Mn^{2+} + 4H_2O$	+1.507
$MnO_4^- + e^- \rightarrow MnO_4^{2-}$	+0.558
$MnO_4^{2-} + 2H_2O + 2e^- \rightarrow MnO_2 + 4OH^-$	+0.60
$Na^+ + e^- \rightarrow Na$	–2.71
$Ni^{2+} + 2e^- \rightarrow Ni$	–0.257
$NiOOH + H_2O + e^- \rightarrow Ni(OH)_2 + OH^-$	+0.49
$NO_3^- + 2H^+ + e^- \rightarrow NO_2 + H_2O$	–0.80
$NO_3^- + 4H^+ + 3e^- \rightarrow NO + 2H_2O$	+0.96
$NO_3^- + H_2O + 2e^- \rightarrow NO_2^- + 2OH^-$	+0.10
$O_2 + 4H^+ + 4\,e^- \rightarrow 2H_2O$	+1.229
$O_2 + e^- \rightarrow O_2^-$	–0.56
$O_2 + H_2O + 2e^- \rightarrow HO_2^- + OH^-$	–0.076
$O_3 + 2H^+ + 2e^- \rightarrow O_2 + H_2O$	+2.076
$O_3 + H_2O + 2e^- \rightarrow O_2 + 2OH^-$	+1.24
$Pb^{2+} + 2e^- \rightarrow Pb$	–0.1262
$Pb^{4+} + 2e^- \rightarrow Pb^{2+}$	+1.67
$PbSO_4 + 2e^- \rightarrow Pb + SO_4^{2-}$	–0.3588
$Pt^{2+} + 2e^- \rightarrow Pt$	+1.118
$Rb^+ + e^- \rightarrow Rb$	–2.98
$S + 2e^- \rightarrow S^{2-}$	–0.4763
$S_2O_8^{2-} + 2e^- \rightarrow 2SO_4^{2-}$	+2.010
$Sn^{2+} + 2e^- \rightarrow Sn$	–0.1375
$Sn^{4+} + 2e^- \rightarrow Sn^{2+}$	+0.151
$Sr^{2+} + 2e^- \rightarrow Sr$	–2.89
$Ti^{2+} + 2e^- \rightarrow Ti$	–1.630
$Ti^{3+} + e^- \rightarrow Ti^{2+}$	–0.368
$Ti^{4+} + e^- \rightarrow Ti^{3+}$	0.00
$U^{3+} + 3e^- \rightarrow U$	–1.798
$U^{4+} + e^- \rightarrow U^{3+}$	–0.607
$V^{2+} + 2e^- \rightarrow V$	–1.175
$V^{3+} + e^- \rightarrow V^{2+}$	–0.255
$Zn^{2+} + 2e^- \rightarrow Zn$	–0.7618
$Zn^{2+} + 2e^- \rightarrow Zn(Hg)$	–0.7628

Table 10.4(a): Buffer solutions and their pH values at 298 K

pH	*Composition of buffer solution*
1.00	25.0 cm^3 of 0.2 M KCl + 67.0 cm^3 0.2 M HCl; mixture diluted to 100 cm^3
2.00	25.0 cm^3 of 0.2 M KCl + 6.5 cm^3 0.2 M HCl; mixture diluted to 100 cm^3
3.00	50.0 cm^3 of 0.1 M potassium hydrogen phthalate + 22.3 cm^3 of 0.1 M HCl; mixture diluted to 100 cm^3
4.00	41.0 cm^3 of 0.2 M acetic acid + 9.0 cm^3 of 0.2 M sodium acetate
5.00	14.75 cm^3 of 0.2 M acetic acid + 35.25 cm^3 of 0.2 M sodium acetate
6.00	9.0 cm^3 of 0.2 M acetic acid + 191.0 cm^3 of 0.2 M sodium acetate
7.00	12.0 cm^3 of 0.05 M borax + 188.0 cm^3 of a solution containing 12.40 g H_3BO_3 and 2.93 g NaCl in 1 dm^3
8.00	11.0 cm^3 of 0.05 M borax + 29.0 cm^3 of a solution containing 12.40 g H_3BO_3 and 2.93 g NaCl in 1 dm^3
9.00	50.0 cm^3 of 0.025 M borax + 4.6 cm^3 of 0.1 M HCl; mixture diluted to 100 cm^3
10.00	50.0 cm^3 of 0.025 M borax + 18.3 cm^3 of 0.1 M NaOH; mixture diluted to 100 cm^3
11.00	50.0 cm^3 of 0.05 M $NaHCO_3$ + 22.7 cm^3 of 0.1 M NaOH; mixture diluted to 100 cm^3
12.00	50.0 cm^3 of Na2HPO_4 + 26.9 cm^3 of 0.1 M NaOH; mixture diluted to 100 cm^3
13.00	25.0 cm^3 of 0.2 M KCl + 66.0 cm^3 of 0.2 M NaOH; mixture diluted to 100 cm^3

Table 10.4(b): Compounds used to prepare buffer solutions and their properties

Compound	*Molecular formula*	*Relative Molar mass*
Potassium hydrogen phthalate		
Potassium tetraoxalate	$KH_3C_4O_8 . 2H_2O$	254.1913
Disodium hydrogen phosphate	Na_2HPO_4	141.9588
Potassium dihydrogen phosphate	KH_2PO_4	136.0852
Disodium tetraborate (borax)	$Na_2B_4O_7 . 10H_2O$	381.367
Sodium carbonate	Na_2CO_3	105.9887
Sodium hydrogen carbonate	$NaHCO_3$	84.0069

Table 10.5: Values of pH for Primary Standard Reference Solutions

Primary Reference Standard	*Temperature / (°C)*					
	25	*30*	*35*	*37*	*40*	*50*
0.05 mol kg^{-1} potassium hydrogen phthalate	4.005	4.011	4.018	4.022	4.027	4.050
0.1 mol kg^{-1} potassium dihydrogen citrate	3.776	3.766	3.759	3.756	3.754	3.749
0.025 mol kg^{-1} disodium hydrogen phosphate + 0.008695 mol kg^{-1} potassium dihydrogen phosphate	6.865	6.853	6.844	6.841	6.838	6.833
0.01 mol kg^{-1} disodium tetraborate	9.180	9.139	9.102	9.088	9.068	9.011
0.025 mol kg^{-1} sodium hydrogen carbonate + 0.025 mol kg^{-1} sodium carbonate	10.012	9.966	9.926	9.910	9.889	9.828

Table 10.6: Dissociation constants of acids in aqueous solutions at 298 K

Acid	*Step*	*pKa*	*Ka*
Acetic		4.75	1.76×10^{-5}
Ammonium ion		9.25	5.6×10^{-10}
o–aminobenzoic		6.97	1.07×10^{-7}
m–aminobenzoic		4.78	1.67×10^{-5}
p–aminobenzoic		4.92	1.2×10^{-5}
Benzoic		4.19	6.46×10^{-5}
Boric		9.14	7.2×10^{-10}
Bromoacetic		2.69	2.05×10^{-3}
Butanoic		4.82	1.5×10^{-5}
Carbonic	1	6.37	4.30×10^{-7}
Carbonic	2	10.25	5.61×10^{-11}
Chloroacetic		2.85	1.40×10^{-3}
Chromic	1	0.74	1.8×10^{-1}
Chromic	2	6.49	3.20×10^{-7}
Citric	1	3.14	7.10×10^{-4}
Citric	2	4.77	1.68×10^{-5}
Citric	3	6.39	4.1×10^{-7}
Formic		3.75	1.8×10^{-4}
Glutaric	1	4.31	4.58×10^{-5}
Glutaric	2	5.41	3.89×10^{-6}
Glycine		9.78	1.67×10^{-10}
Hydroiodic		−11	10^{11}
Hydrobromic		−9	10^{9}
Hydrochloric		−7	10^{7}

Hydrofluoric		3.45	3.53×10^{-4}
Hydrocyanic		9.31	4.93×10^{-10}
Hydronium ion		0.0	1
Hybromous		8.69	2.60×10^{-9}
Hypochlorous		7.53	3.0×10^{-8}
Hypoiodous		10.64	2.3×10^{-11}
Lactic		3.86	1.4×10^{-4}
Maleic	1	1.83	1.42×10^{-2}
Maleic	2	6.07	8.57×10^{-7}
Malic	1	3.40	3.9×10^{-4}
Malic	2	5.11	7.8×10^{-6}
Malonic	1	2.83	1.49×10^{-2}
Malonic	2	5.69	2.03×10^{-6}
o–nitrobenzoic (at 291 K)		2.16	6.95×10^{-3}
m–nitrobenzoic		3.47	3.4×10^{-4}
p–nitrobenzoic		3.41	3.93×10^{-4}
Oxalic	1	1.23	5.90×10^{-2}
Oxalic	2	4.19	6.40×10^{-5}
Perchloric		–10	10^{10}
Periodic		1.64	2.3×10^{-3}
Phenol		9.89	1.3×10^{-10}
o–Phosphoric	1	2.12	7.52×10^{-3}
o–Phosphoric	2	7.21	6.23×10^{-8}
o–Phosphoric	3	12.67	2.2×10^{-13}
Picric		0.38	4.2×10^{-1}
Propanoic		4.87	1.34×10^{-5}
Succinic	1	4.16	6.89×10^{-5}
Succinic	2	5.61	2.47×10^{-6}
Sulphuric	1	–2	10^{2}
Sulphuric	2	1.92	1.2×10^{-2}

Table 10.7: Data on sparingly soluble salts of silver, mercury and lead commonly used in precipitation potentiometric titrations All values at 298 K

Salt	*Half–cell reaction*	*Standard potential E^0 / (V)*	*Solubility product*
AgI	$AgI + e^- \rightarrow Ag + I^-$	–0.15224	8.51×10^{-17}
AgCN	$AgCN + e^- \rightarrow Ag + CN^-$	–0.017	5.97×10^{-17}
AgBr	$AgBr + e^- \rightarrow Ag + Br^-$	+0.07133	5.35×10^{-13}
AgSCN	$AgSCN + e^- \rightarrow Ag + SCN^-$	+0.08951	1.03×10^{-12}
AgCl	$AgCl + e^- \rightarrow Ag + Cl^-$	+0.02223	1.77×10^{-10}
$AgIO_3$	$AgIO_3 + e^- \rightarrow Ag + IO_3^-$	+0.354	3.17×10^{-8}
$AgBrO_3$	$AgBrO_3 + e^- \rightarrow Ag + BrO_3^-$	+0.546	5.34×10^{-5}

$Ag_2C_2O_4$	$Ag_2C_2O_4 + 2e^- \rightarrow Ag + C_2O_4^{2-}$	+0.4647	5.40×10^{-12}
Ag_2CrO_4	$Ag_2CrO_4 + 2e^- \rightarrow Ag + CrO_4^{2-}$	+0.4470	1.12×10^{-12}
Ag_2O	$Ag_2O + H_2O + 2e^- \rightarrow 2Ag + 2OH^-$	+0.342	2.0×10^{-8}
PbI_2	$PbI_2 + 2e^- \rightarrow Pb + 2I^-$	−0.365	8.49×10^{-9}
$PbBr_2$	$PbBr_2 + 2e^- \rightarrow Pb + 2Br^-$	−0.284	6.60×10^{-6}
$PbCl_2$	$PbCl_2 + 2e^- \rightarrow Pb + 2Cl^-$	−0.2675	1.17×10^{-5}
$PbSO_4$	$PbSO_4 + 2e^- \rightarrow Pb + SO_4^{2-}$	−0.3588	1.82×10^{-8}
$Pb(IO_3)_2$	$Pb(IO_3)_2 + 2e^- \rightarrow Pb + 2IO_3^-$	−0.496	3.68×10^{-13}
$PbCrO_4$	$PbCrO_4 + 2e^- \rightarrow Pb + CrO_4^{2-}$	−0.530	1.8×10^{-14}
Hg_2I_2	$Hg_2I_2 + 2e^- \rightarrow 2Hg + 2I^-$	−0.0405	5.33×10^{-29}
Hg_2Br_2	$Hg_2Br_2 + 2e^- \rightarrow 2Hg + 2Br^-$	+0.1392	6.41×10^{-23}
Hg_2Cl_2	$Hg_2Cl_2 + 2e^- \rightarrow 2Hg + 2Cl^-$	+0.2681	1.45×10^{-18}
$Hg_2(SCN)_2$	$Hg_2(SCN)_2 + 2e^- \rightarrow 2Hg + 2SCN^-$	−0.22	3.12×10^{-20}
$Hg_2(CN)_2$	$Hg_2(CN)_2 + 2e^- \rightarrow 2Hg + 2CN^-$	+0.36	5×10^{-40}
$Hg_2C_2O_4$	$Hg_2C_2O_4 + 2e^- \rightarrow 2Hg + C_2O_4^{2-}$	−0.41	1.75×10^{-13}
Hg_2CrO_4	$Hg_2CrO_4 + 2e^- \rightarrow 2Hg + CrO_4^{2-}$	−0.54	2×10^{-9}

Expected Experimental Results

Experiment 10.1

Refer to table 10.7.

Experiment 10.2

Refer to table 10.6.

Experiment 10.3

Refer to table 10.6.

Experiment 10.5

$[Ag(NH_3)_2]^+$.

Experiment 10.6

The equilibrium constant is of the order of 10^{-7}.

Experiment 10.7

$pK_1 = 2.34, pK_2 = 9.60, pK_1 = 5.97$

Experiment 10.8

pK values at 25 °C:

- benzoic acid = 4.20
- *p*-nitrobenzoic acid = 3.42
- *p*-aminobenzoic acid = 4.92

Hammett acidity constants at 25°C:

- *p*-nitrobenzoic acid = 0.47
- *p*-aminobenzoic acid = –0.44

Hints to Review Questions

1. See Section 10.7.6
2. See Section 10.7.6
3. Did you guess pH 9? Common sense dictates that a solution of pH 9 would be alkaline in nature so a solution of hydrochloric acid cannot have pH more than 7. At concentrations as low as 10^{-9} M HCl, the H^+ ions produced due to dissociation of water molecules also become important. Pure water has an ionic product of 10^{-14} which means that it is 10^{-7} M in H^+ ions.
   Total H^+ ion concentration in solution = 10^{-7} (from water) + 10^{-9} (from HCl)
   $$= 101 \times 10^{-9} \text{ M}$$
   $$\text{pH} = -\log [H^+] = -\log 101 \times 10^{-7} \approx 6.9$$
4. Is your answer zero? The pH scale, as devised by Sorenson, ranges from 1 to 14 and is not adequate to handle hydrogen ion concentration as high as 1 M.
5. When the reactions at the electrodes of the two half-cells are identical, the cell is called a concentration cell. See Section 10.10.

Further Exploration through the Internet

http://ull.chemistry.uakron.edu/analytical/Potentiometry/
http://www.chem.vt.edu/chem-ed/echem/potentio.html
http://ibc.wsu.edu/faculty/dkother/course2001/021000/Redox.htm
http://www.life.uiuc.edu/crofts/bioph354/redox.html
http://courses.chem.ku.edu/Fall99/516_Fall99a/Potentiometry.html
http://www.trincoll.edu/~henderso/textfi~1/Chem%20208%20notes/potentiometry.htm
http://www.uvm.edu/~jgoldber/courses/chem121/prob5.html
http://www.chemeng.drexel.edu/web_books/EngBio/Hidden/sensr/ch2/2_4f.htm
http://www.auburn.edu/~cammavi/class/chem6130/potent/
http://www.chem.fsu.edu/tvickers/31echem/potentiometry.htm

Jaroslav Heyrovsky

Jaroslav Heyrovsky was born in 1890 in Prague, Czechoslovakia. He took his B.Sc. degree in 1913 from University College, London, working under Professors Sir William Ramsay and F. G. Donnan. At Professor Donnan's suggestion, he began studies of liquid metal electrodes (aluminium amalgams) that were continuously renewed from glass capillaries.

In 1914 World War I erupted and Heyrovsky was called back from England to serve in the Austro-Hungarian army. He worked in a military hospital as a dispensing chemist and radiologist. He continued his experiments in the hospital pharmacy and was able to prepare a Ph.D. dissertation while still a soldier. In 1918 the war ended and he appeared for his thesis defence at the Prague University. There he met Professor Kucera who invited Heyrovsky to join his research group, which was working on electrocapillary measurements of the Dropping Mercury Electrode (DME).

Heyrovsky's initial experiment consisted of applying a voltage to the DME and immersing a reference electrode in the test solution. After 50 drops of mercury were collected, they were dried and weighed. The applied voltage was varied and the experiment repeated. A graph of measured weight versus applied voltage was plotted. Heyrovsky cut down on the tediousness of the weighing step by monitoring the drop-time instead. In 1921 he eliminated this step too by measuring the current flow through the cell. His first measurement was made using an amperometer. It didn't work. Undaunted, he used a sensitive galvanometer instead. On February 10, 1922 recorded the current-voltage curve for a solution of 1 M NaOH and correctly interpreted the rise in current between – 1.9 V and – 2.0 V as being due to the reduction of sodium ions at the DME.

The polarograph was born. Thereafter, Heyrovsky concentrated his entire scientific research on the development of this new branch of electrochemistry. In 1959, he was awarded the Nobel Prize for his discovery and development of polarography. He was director of the Polarography Institute at the Czechoslovak Academy of Sciences till 1963. He died in 1967 at Prague.

11 Voltammetry

11.1 Introduction

Voltammetry is a group of electrochemical techniques that are widely used in analytical chemistry for the detection and identification of organic and inorganic species present in trace amounts in small samples. The oldest and the most widely used voltammetric technique is *polarography*. The technique was modified to give rise to several others like current-sampled polarography, pulse polarography and differential pulse polarography. Modern voltammetric techniques also include fast linear-sweep polarography, cyclic voltammetry, alternating-current voltammetry, and stripping voltammetry.

11.2 Terminology

Voltammetry is the study of how the current flowing during electrolysis in a cell changes with the applied voltage. The electrode at which the analyte ions are discharged is known as the *working electrode*. It is commonly made up of gold, platinum or graphite. When a Dropping Mercury Electrode (DME) is used as the working electrode, the technique is known as *polarography*.

To understand the phenomenon of voltammetry, several other concepts need to be mastered first.

11.2.1 Why do ions travel to electrodes?

Consider a solution in which two electrodes dip. When a potential is applied across the electrodes, the ions in the solution move towards the electrodes. This movement is known as *mass transfer* of ions. There are three processes that cause the mass transfer of ions to and from the electrodes:

- *Electrostatic attraction* between an ion and an oppositely charged electrode. The current that results from this process is called *migration current*.
- *Mechanical movement* caused by stirring of the solution, thermal convection currents, density gradients within the solution, and vibration of the apparatus.
- *Diffusion* due to a concentration gradient near the electrode surface.

In voltammetry, migration current and convection current are suppressed so that the entire flow of current is due to the process of diffusion alone. Migration current can be eliminated by adding a *supporting electrolyte* to the analyte such that its concentration is at least a hundred times that of the analyte ion. The ions of the supporting electrolyte migrate to the electrode due electrostatic attraction but cannot be discharged there because the required electrode potential is absent. They remain in the vicinity of the electrode and thus serve to shield the analyte ions in the bulk of the solution from the charge on the electrode. Convection currents are suppressed by damping vibrations in the apparatus and maintaining a uniform temperature throughout the experiment.

11.2.2 Polarised and non-polarised electrodes

Non-polarised electrodes are those whose electrode potential does not depend upon the magnitude of current flowing though the cell. Reference electrodes used in potentiometric measurements (see Potentiometry, Section 10.7) are examples of non-polarisable electrodes. It follows that a *polarisable electrode* is one whose potential is easily changed by the flow of even a tiny current.

Polarisation is the general phenomenon of deviation in the potential of an electrode from any steady, open-circuit value, brought about by the flow of current.

Electrodes with a very small surface area (of the order of a few square millimetres) tend to get polarised more easily. They are generally called *microelectrodes*.

In voltammetry, polarisable microelectrodes serve as the electrode at which the analyte ions are discharged.

11.2.3 Overpotential

Consider an electrolysis cell in which two smooth platinum electrodes are dipping in an electrolyte of 1 M H_2SO_4. An external voltage is applied across the two electrodes. When the applied voltage is very small, say 0.5 V, the current flowing through the cell is tiny. As the applied voltage is increased, the current rises slightly until, at about 1.7 V, it shoots up. At this point bubbles of gas are liberated at the two electrodes; hydrogen gas is evolved at the cathode and oxygen at the anode. The net reaction is evidently the decomposition of water.

Reaction at anode: $2H^+ + 2e^- \rightarrow H_2$ (11.1)

Reaction at cathode: $2H_2O \rightarrow O_2 + 4H^+ + 4e^-$ (11.2)

Net reaction: $2H_2O \rightarrow 2H_2 + O_2$ (11.3)

Thus 1.7 V is the approximate *decomposition voltage* of water. However, the reversible emf of the cell calculated from the data given in Table 10.1 in Appendix 10 has a value of 1.23 V. The difference of approximately 0.5 V between the value and the observed decomposition voltage (1.7 V) is known as the *overvoltage* required to force the reaction of equation (11.3) to proceed at an appreciable rate.

Overvoltage can thus be defined as the voltage in excess of the reversible emf that must be applied to a cell to cause the cell reaction to proceed at an appreciable rate. The overvoltage of a cell is the sum of the *overpotentials* at the cathode and the anode. (The terms 'overvoltage' and 'overpotential' are analogous, the only difference being that the former is applied to a cell and the latter to a single electrode.)

The overpotential at an electrode depends on the following factors:

- *Current density*: Current density is the current in ampere at the electrode per unit area of the electrode. Overpotential increases rapidly with increasing current density up to 0.01 A cm^{-2}. Above this value the increase is less rapid.
- *The nature and physical state of the metal used as the electrode*. The hydrogen overpotential at a platinised platinum electrode is much less than that at a smooth one because the platinised electrode has a larger surface area and hence smaller current density.
- *The physical state of the substance deposited at the cathode.* Overpotentials for gases are usually much larger than those for metals.
- *Temperature.* As the temperature increases, the overpotential usually decreases to a great extent.
- *Concentration gradient at the electrode-solution interface.* The overpotential increases with increasing concentration gradient. The concentration gradient itself depends on the rate of stirring of the solution, the current density and the temperature.

Polarography employs mercury to form the electrode at which discharge of the analyte ions takes place. Hydrogen has a very high overpotential on mercury (approximately 0.8 V in 1 M H_2SO_4). Thus the range of cathodic potentials over which the mercury electrode can function becomes greatly extended. Species whose reduction potential is considerably more negative than the reversible potential for the discharge of hydrogen can be discharged at the mercury electrode without interference from hydrogen ion formation.

11.2.4 Faradaic and non-faradaic currents

Redox reactions that take place at electrodes involve the transfer of electrons between the electrode and the solution surrounding it. They are also known as *charge-transfer reactions*. Such reactions obey Faraday's laws of electrolysis (see Section 9.2). Hence such reactions are *faradaic* in nature, and the current generated due to these reactions is known as *faradaic current*.

It is possible for conditions to occur such that a current is generated even in the absence of any charge-transfer reaction. Absorption or desorption on the electrode surface can lead to a change in the electrode-solution interface when changes in potential or surface composition occur. Such processes are *non-faradaic* in nature and generate *non-faradaic currents*.

11.3 The Dropping Mercury Electrode (DME)

The DME (shown in Figure 11.1) is the polarisable microelectrode used in polarography. The electrode surface is the surface of a fine mercury drop produced by forcing pure, liquid mercury through a narrow glass capillary (0.05 – 0.08 mm internal diameter and 5 – 9 cm length) under a pressure head of about 50 cm. A succession of fine mercury drops is formed at the rate of one every 2 – 6 seconds. Thus every 2 – 6 seconds, the analyte reduces at a *new* electrode surface. The renewable electrode surface provided by the DME is one of its most outstanding features. A constant drop time is ensured by using mechanical knockers to detach the drop from the capillary. As the detached drop falls through the analyte solution, the analyte ions get reduced at the surface of the drop.

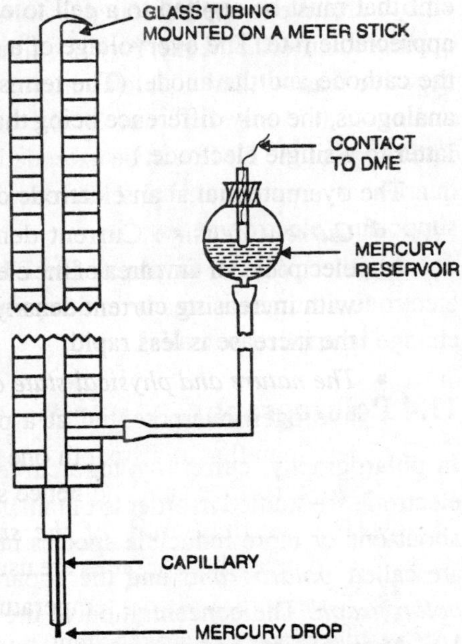

Figure 11.1: The Dropping Mercury Electrode.

Compared to the electrodes generally used in potentiometry (see Section 10.7) the DME is unusual. What advantages does the DME have over these electrodes? Electrodes used in potentiometry have only one electrode surface throughout their life. During use, it is possible that this surface is attacked by a species that reduces its effectiveness by poisoning or passivating it. The DME avoids this by exposing a clean electrode surface to the analyte solution every time a new mercury drop is formed. The surface area of a mercury drop produced from a capillary is quite reproducible so it is reasonable to assume that all successive mercury drops are identical.

Another characteristic of electrodes used in potentiometry is that they are generally made of conducting materials that do not react with the analyte ions being discharged on their surface. The most common examples are gold, platinum, and graphite. Mercury, on the other hand, forms amalgams with many metals, which lowers their reduction potential. Most analyte solutions are aqueous in nature and thus suffer the risk of interference from hydrogen ion discharge at the electrode. However, hydrogen has a high overpotential on mercury. This makes it possible to discharge species whose reduction potential is more negative than the reversible potential for the discharge of hydrogen, e.g., alkali metal ions, Al(III) ion, Mn(II) ion, etc.

The range over which the DME is used in aqueous solutions is + 0.3 to –2.0 V relative to the Saturated Calomel Electrode (SCE). At potential more positive than + 0.3 V, mercury is oxidised to mercury (I). At potentials more negative than – 1.2 V visible hydrogen evolution occurs in 1 M HCl solutions and at – 2.0 V the alkali ions of the supporting electrolyte start discharging. The range of the DME can be extended to about – 2.6 V versus SCE by using quarternary ammonium salts, such as tetra-alkylammonium hydroxide, as the supporting electrolyte.

Another point in favour of the DME is that the diffusion current at the electrode assumes a steady and reproducible value immediately after each change in applied potential.

11.4 Polarography

In polarography, current-voltage curves obtained with a dropping mercury electrode are studied in order to obtain quantitative and qualitative information about one or more reducible species in solution. The current-voltage curves are called *polarograms* and the apparatus used to obtain them is called a *polarograph*. The concentration of the electroactive species can range from 10^{-2} M to 10^{-5} M, and the volume of the sample solution from 2 to 25 cm^3.

11.4.1 Experimental set-up

11.4.1.1 The DME

The mercury used for the DME should be extremely pure. Purification can be done by double distillation or by filtering re-distilled mercury through a sintered-glass funnel. The capillary of the DME should never be allowed to get clogged. It can be cleaned from the inside by passing concentrated nitric acid through it till the blocking matter has dissolved. Distilled water is drawn through the capillary to wash off the acid completely. Warm air, filtered through cotton wool, is passed through the capillary to dry it.

The DME assembly is mounted on a heavy support stand on a vibration-free surface. It should be mounted within ± 5°C of the vertical otherwise the mercury drops do not fall in a uniform, periodic manner. Before using the DME, the capillary tip should be immersed in 1:1 nitric acid while the mercury is flowing. The tip is then washed well with distilled water.

11.4.1.2 Polarographic cells

The original cell used by Heyrovsky consisted of an Erlenmeyer flask (Figure 11.2). It contained a large pool of mercury at the bottom that served as the anode. Due to its large surface area, the mercury pool is incapable of becoming polarised. Its potential depends on the nature and concentration of the supporting electrolyte. When the analyte solution contains ions that are capable

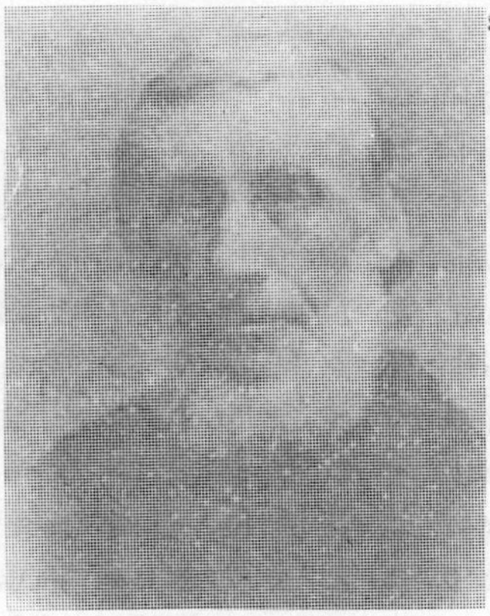

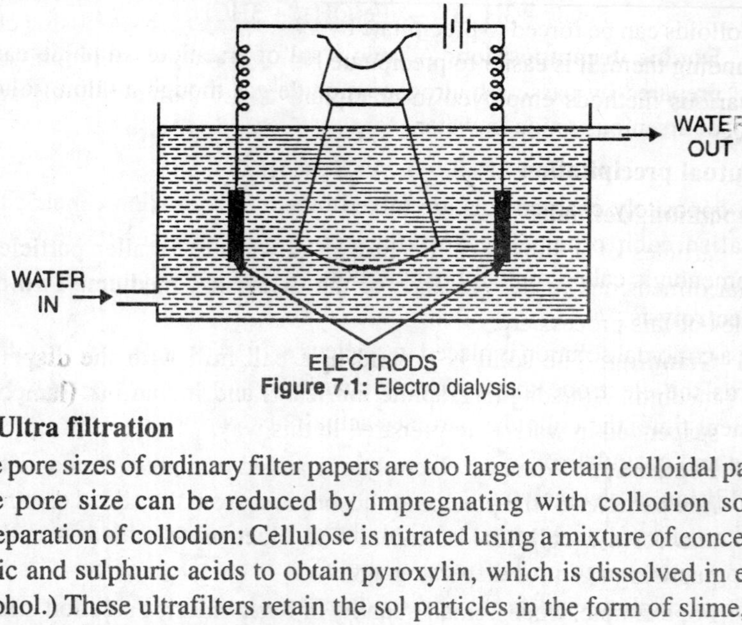

Element	*Ion electrode reaction*	*Supporting Electrolyte*	$E_{1/2}$ /(*volts versus SCE)*
Cobalt	$Co^{2+} \rightarrow Co$	0.05 M K_2SO_4	- 1.428
	$Co^{2+} \rightarrow Co$	0.1 M KCl	- 1.20
	$Co^{2+} \rightarrow Co$	0.1 M pyridine + 0.1 M pyridinium chloride	- 1.07
	$Co^{2+} \rightarrow Co$	1 M KSCN	- 1.03
	$Co^{3+} \rightarrow Co$		
	$Co^{2+} \rightarrow Co(Hg)$	1 M NH_3 + 1 M NH_4Cl	- 0.5 (first wave) - 1.3 (second wave)
Chromium	$Cr^{3+} \rightarrow Cr^{2+}$ $Cr^{2+} \rightarrow Cr$	0.1 M KCl	-0.91 (first wave) -1.47 (second wave)
	$Cr^{3+} \rightarrow Cr^{2+}$	0.1 M pyridine + 0.1 M pyridinium chloride	- 0.95
	$Cr^{2+} \rightarrow Cr^{3+}$	1 M KCl	- 0.40
	$Cr^{2+} \rightarrow Cr^{3+}$	1 M KSCN	- 0.80
Caesium	$Cs^{+} \rightarrow Cs$	0.1 M $N(C_2H_5)_4OH$ in 50% C_2H_5OH	- 2.05
Copper	$Cu^{2+} \rightarrow Cu$	0.1 M KCl (HCl)	+ 0.04
	$Cu^{2+} \rightarrow Cu$	1 M NaOH	- 0.42
	$Cu^{2+} \rightarrow Cu$	0.5 M sodium tatrate, pH 9	- 0.12
	$Cu^{2+} \rightarrow Cu^{+}$	1 M $K_2C_2O_4$, pH 7.5 – 10	- 0.27
	$Cu^{2+} \rightarrow Cu^{+}$ $Cu^{+} \rightarrow Cu$	1 M NH_3 + 1 M NH_4Cl	- 0.24 (first wave) - 0.50(second wave)
Iron	$Fe^{3+} \rightarrow Fe^{2+}$	0.5 M sodium citrate, pH 6	- 0.222
	$Fe^{3+} \rightarrow Fe^{2+}$ $Fe^{2+} \rightarrow Fe$	0.5 M sodium citrate, pH 5.8	- 0.21 (fisrt wave) - 1.53(second wave)
	$Fe^{3+} \rightarrow Fe^{2+}$ $Fe^{2+} \rightarrow Fe$	0.5 M sodium tartrate, pH 9.4	- 1.20 (first wave) - 1.73(second wave)
	$Fe^{3+} \rightarrow Fe^{2+}$ $Fe^{2+} \rightarrow Fe$	0.1 M EDTA + 2 M CH_3COONa	- 0.13 (first wave) - 1.3(second wave)
	$Fe^{3+} \rightarrow Fe^{2+}$	0.2 M $Na_2C_2O_4$, pH 5.25	- 0.245
	$Fe^{2+} \rightarrow Fe$	1 M NaOH	- 1.46
	$Fe^{2+} \rightarrow Fe$	0.1 M KCl	- 1.3

Element	*Ion electrode reaction*	*Supporting Electrolyte*	$E_{1/2}$ /(*volts versus SCE*)
Lead	$Pb^{2+} \rightarrow Pb$	0.1 M KCl	- 0.396
	$Pb^{2+} \rightarrow Pb$	1 M HNO_3	- 0.405
	$Pb^{2+} \rightarrow Pb$	1 M NaOH	- 0.755
	$Pb^{2+} \rightarrow Pb$	0.5 tartrate + 0.1 M NaOH	- 0.75
	$Pb^{2+} \rightarrow Pb$	0.5 M sodium hydrogen tartrate, pH 4.5	- 0.48
	$Pb^{2+} \rightarrow Pb$	1 M KCN	- 0.72
Lithium	$Li+ \rightarrow Li(Hg)$	0.1 M $N(C_2H_5)_4OH$ in 50% C_2H_5OH	- 2.31
Magnesium	$Mg^{2+} \rightarrow Mg(Hg)$	0.1 M $N(CH_3)_4Cl$	- 2.2
Manganese	$Mn^{2+} \rightarrow Mn$	1 M KCl	- 1.51
	$Mn^{2+} \rightarrow Mn$	1 M NH_3 + 1 M NH_4Cl	- 1.65
	$Mn^{2+} \rightarrow Mn$	1.5 M KCN	- 1.33
Nickel	$Ni^{2+} \rightarrow Ni$	0.01 M KCl	- 1.1
	$Ni^{2+} \rightarrow Ni$	1 M KSCN	- 0.70
	$Ni^{2+} \rightarrow Ni$	1 M KCl + 0.5 M pyridine, pH 7.0	- 0.78
	$Ni^{2+} \rightarrow Ni$	1 M NH_3 + 1 M NH_4Cl	- 1.09
	$Ni^{2+} \rightarrow Ni$	1 M KCN	- 1.36
Oxygen	$O_2 \rightarrow H_2O_2$		- 0.05 (first wave)
	$H_2O_2 \rightarrow H_2O$	pH 1 – 10	-0.94(second wave)
Palladium	$Pd^{2+} \rightarrow Pd$	1 M KCl + 1 M pyridine	- 0.36
	$Pd^{2+} \rightarrow Pd$	1 M KCN	- 1.77
	$Pd^{2+} \rightarrow Pd$	1 M NH_3 + 1 M NH_4Cl	- 0.80
Potassium	$K^+ \rightarrow K(Hg)$	0.1 M $N(C_2H_5)_4OH$ in 50% C_2H_5OH	- 2.10
Sodium	$Na^+ \rightarrow Na(Hg)$	0.1 M $N(C_2H_5)_4OH$ in 50% C_2H_5OH	- 2.07
Strontium	$Sr^{2+} \rightarrow Sr(Hg)$	0.1 M $N(C_2H_5)_4I$ in 50% C_2H_5OH	- 2.06
Tin	$Sn^{2+} \rightarrow Sn$	1 M HCl	- 0.47
	$Sn^{2+} \rightarrow Sn$		- 1.22 (fisrt wave)
	$Sn^{2+} \rightarrow Sn^{4+}$	1 M NaOH	- 0.93(second wave)
	$Sn^{4+} \rightarrow Sn^{2+}$		- 0.25 (first wave)
	$Sn^{2+} \rightarrow Sn$	4 M NH_4Cl + 1 M HCl	- 0.52(second wave)

Element	*Ion electrode reaction*	*Supporting Electrolyte*	*$E_{1/2}$ /(volts versus SCE)*
Thallium	$Tl^{+} \rightarrow Tl$	0.1 M KNO_3 or 0.1 M KCl	-0.460
Titanium	$Ti^{4+} \rightarrow Ti^{2+}$ $Ti^{2+} \rightarrow Ti^{4+}$	0.1 M HCl 0.01 M HCl	- 0.81 - 0.14
Zinc	$Zn^{2+} \rightarrow Zn$ $Zn^{2+} \rightarrow Zn$ $Zn^{2+} \rightarrow Zn$	0.1 M KCl 1 M NaOH 1 M NH_3 + 1 M NH_4Cl	- 0.995 - 1.53 - 1.33

EXPECTED RESULTS

Experiment 11.1

Cu(II) = + 0.04 V, Cd(II) = - 0.64 V; Zn(II) = -0.995 V (all versus SCE)

Experiment 11.2

$p = 2, K = 5 \times 10^6$

Experiment 11.4

Formal redox potential of the $K_3[Fe(CN)_6]/K_4[Fe(CN)_6]$ couple = 0.116 V versus SCE

Number of electrons involved in the redox reaction = 1.

The ratio of the cathodic peak current to the anodic peak current should be unity for a redox reaction to be reversible.

(Deviations from the expected values could occur because nitrogen gas has not been passed through the solution long enough, the temperature has not been maintained constant or the surface area of the platinum electrode was too large.)

Hints to Review Questions

1. Look at the discussion on analysis copper alloys in Section 11.4.3.6.
2. In cyclic voltammetry, the potential is first increased linearly to a peak and then decreased to its original value at the same rate. The plot of potential versus time is of the form of a triangle and hence the name.
3. Residual current is eliminated by subtracting the value of the residual current from the corresponding value of the limiting current to obtain the correct value of the diffusion current.

Further Exploration through the Internet

http://ull.chemistry.uakron.edu/analytical/Voltammetry/

http://www.auburn.edu/~cammavi/class/chem6130/voltam/

http://www.epsilon-web.net/Ec/manual/EC_FAQs/faqvolt.html

http://www.siue.edu/~michsha/c549/chem549a.htm

http://www.compusmart.ab.ca/plambeck/che/p415/p15135.htm

http://elchem.kaist.ac.kr/vt/chem-ed/echem/polarogr.htm

http://www.chemistry.adelaide.edu.au/external/Soc-Rel/Content/polarogr.htm

http://www.chem.usu.edu/faculty/SBialkow/Classes/565/polar.html

http://www-biol.paisley.ac.uk/marco/Enzyme_Electrode/Chapter1/Cyclic_Voltammetry1.htm

http://soulcatcher.chem.umass.edu/web/downloads/cyclicv.html

Robert Wilhelm Bunsen

Robert Wilhelm Bunsen was born on March 31, 1811 in Guttingen, Germany. He took his Ph.D. in chemistry at the University of Guttingen in 1830. Later as a lecturer there he investigated the solubility of arsenic salts. In 1834 he found that freshly precipitated hydrated iron (III) oxide acts as an antidote to arsenic poisoning, still the best known antidote today.

Bunsen later taught at the University of Marsburg where he conducted experiments with the highly toxic, arsenic-containing compound cacodyl. Though this work made him famous quickly, it also nearly killed him by arsenic poisoning and made him lose the sight in one eye when the compound exploded and sent a sliver of glass into his eye. His student, Edward Frankland, went on to successfully study more organometallic compounds.

While at Marsburg, Bunsen showed that the charcoal-burning furnaces in Germany were extremely inefficient. He suggested techniques that could recycle gases and recover valuable by-products like ammonia. In 1841 he created the "Bunsen battery" which used carbon as an electrode instead of expensive platinum. This cheap battery was used in the large-scale production of arc light and in electroplating.

In 1852 Bunsen was appointed as a professor at the University of Heidelberg. There he built up a laboratory of such excellence that it attracted students and chemists from all over the world. He continued his research in batteries. By substituting nitric acid with chromic acid, he was able to produce pure metals such as Li, Na, Mg, Al, Ca, Ba, Cr and Mn by electrolysis. Being the first to obtain magnesium in the metallic state, he studied its physical and chemical properties. The brilliant white light produced when magnesium burns triggered Bunsen's interest in photochemistry. He studied the formation of HCl from hydrogen and chlorine gas in the presence of light.

In 1859, Bunsen became excited by Kirchoff's discovery that the emission spectra of elements excited by a burner flame are better resolved by using a prism. Bunsen quickly found that the higher the temperature and the lower the luminosity of the flame itself, the better the spectra. He modified the existing design of the burner to premix the gas and air prior to combustion. The newly designed burner was named after him.

Bunsen and Kirchoff studied the spectral lines of a number of elements. Starting from a prism, a cigar box and the ends of two old telescopes, they developed a spectroscope, thus giving rise to the field of spectroscopy. Using this instrument, they discovered two new elements: caesium and rubidium. They were named for the colour of the spectral lines they produced (Latin *caesium*, sky blue; *rubidum*, darkest red colour).

Bunsen retired from his chemical work at the ripe old age of 78. He died on August 16, 1899.

12 Spectroscopy

12.1 Introduction

Spectroscopy is the study of the interaction of light with matter. Light can be absorbed, scattered or emitted by atoms, ions and molecules. The result of this interaction is a change in the energy state of the chemical species. Since each chemical species has its own set of characteristic energy states, spectroscopy allows us to delve right into the world of atoms and molecules. Before exploring this exotic world, it is worthwhile understanding what is light.

12.2 The Nature of Light

The nature of light has puzzled scientists for over two hundred years, often dividing them into two groups: those who argued that light has wave nature and those who believed that light behaved like a particle. The controversy was finally resolved when it was realised that light has *dual nature*; it is capable of behaving as both a wave and a particle. However, it is impossible to measure simultaneously and accurately the wave and particle properties of light, a statement known as the Heisenberg Uncertainty Principle. If an experiment is set-up that detects waves then light will show wave properties; if an experiment is set up that detects particles then light will exhibit particle properties.

In 1927 it was shown that even matter, usually thought of as consisting of particles, could exhibit wave-like properties. Two American physicists, Davisson and Germer, performed an experiment that proved that electrons could be diffracted. Diffraction is a phenomenon exhibited by waves. This remarkable dual behaviour had already been predicted in 1925 by a Frenchman, Prince Louis de Broglie. He had put forward an equation relating the particle-like property of momentum, mv, with the wave-like property of wavelength, λ.

$$mv = \frac{h}{\lambda} \tag{12.1}$$

where h is the Planck constant.

This equation is known as the *de Broglie equation.*

Sir J. J. Thomson was awarded the Nobel Prize in 1906 for his experimental work proving the particle-like properties of the electron. In one of those remarkable coincidences that abound in the history of science, Sir Thomson's son, G. P. Thomson was awarded the Nobel Prize in 1937 for his work that proved that the electron has wave-like properties.

12.2.1 The wave nature of light

A light wave is an electromagnetic wave, i.e., it has an electric field and a magnetic field associated with it. The electric and magnetic fields are always at right angles to each other and to the direction in which the wave is travelling. For example, if the light wave is travelling in the X direction then the electric field oscillates in the XZ plane and the magnetic field in the YZ plane, as shown in Figure 12.1.

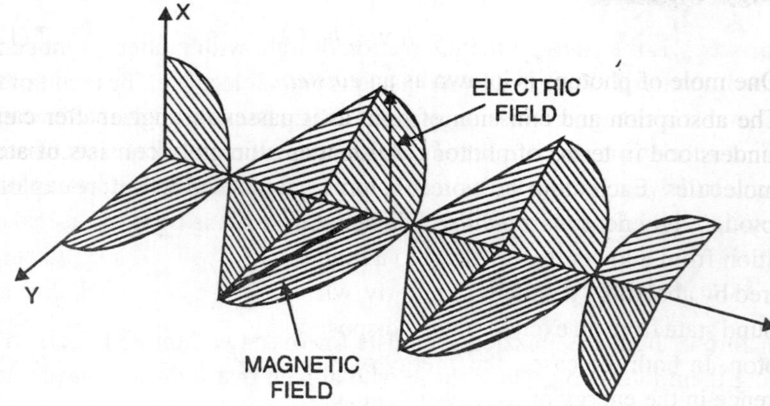

Figure 12.1: The electric and magnetic fields associated with light.

Light travels with a velocity of approximately 3×10^8 m s^{-1} in vacuum. This velocity decreases when light traverses a medium. The ratio of the velocity of light in vacuum to that in a particular medium is known as the *refractive index* of that medium. A wave can be characterised by its *wavelength* and *frequency*. A wave consists of crests and troughs, as shown in Figure 12.2. The distance between and two consecutive crests or any two consecutive troughs is the *wavelength*. The number of waves that pass a fixed point in one second is the *frequency*. The wavelength is related to the frequency by the equation

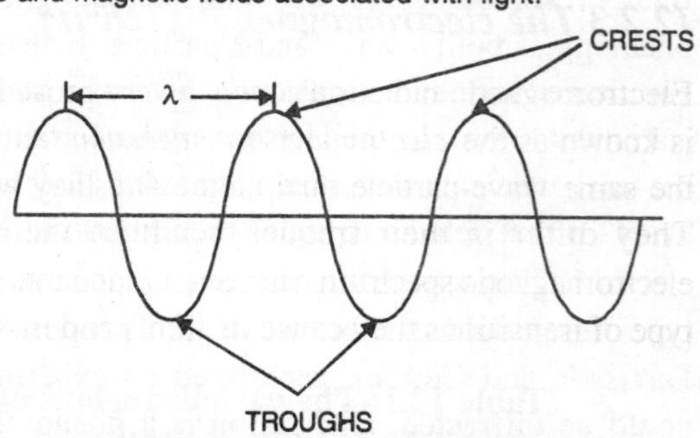

Figure 12.2: The concept of wavelength.

$$c = \nu\lambda \tag{12.2}$$

where c is the velocity of light.

The frequency of light remains unchanged when it propagates through any medium.

The *wavenumber* $\bar{\nu}$ is the reciprocal of the wavelength.

$$\bar{\nu} = 1/\lambda \tag{12.3}$$

The SI unit of wavelength is metre. It follows that the SI unit of wavenumber is m^{-1} but units of cm^{-1} are most commonly used. Frequency is expressed in hertz (Hz).

12.2.2 The particle nature of light

The particles that make up light are known as *photons.* Since light is a form of energy a photon can be thought of as a small packet of energy known as *quantum* of energy. The energy, *E*, of a photon depends on its frequency.

$$E = h\nu = hc/\lambda \qquad (12.4)$$

One mole of photons is known as an *einstein.*

The absorption and emission of light as it passes through matter can be best understood in terms of photons. All matter ultimately consists of atoms and molecules. Each atom or molecule has a number of energy states it can occupy, i.e., its energy is *quantised.* When an atom or a molecule makes a transition from a lower energy level to a higher one, it gains the extra energy required by absorbing a photon. Similarly, when an atom or a molecule comes to ground state from an excited state, it disposes of the extra energy by emitting a photon. In both the cases, the energy of the photon is exactly equal to the difference in the energy of the lower state and the higher state.

12.2.3 The electromagnetic spectrum

Electromagnetic radiation covers a very broad range of frequencies. This range is known as the *electromagnetic spectrum.* All radiations in the spectrum have the same wave-particle dual nature and they all travel with the speed of light. They differ in their frequency and in their interaction with matter. The electromagnetic spectrum has been divided into a number of regions based on the type of transitions they cause in atoms and molecules, as shown in Table 12.1.

Table 12.1: The various regions of the electromagnetic spectrum and the transitions associated with them

Region in electro-magnetic spectrum	Wavelength λ / (m)	Frequency ν / (Hz)	Energy *E* / (J mol^{-1})	Transitions caused	Information gained
Gamma rays	$10^{-10} - 10^{-12}$	$10^{18} - 10^{20}$	10^{9} to 10^{12}	Transitions in energy levels of a nucleus	Structure of the nucleus
X-Rays	$10^{-9} - 10^{-10}$	$10^{17} - 10^{18}$	$10^{7} - 10^{9}$	Transitions of inner shell electrons of an atom	Electronic structure of an atom

Region in electro-magnetic spectrum	Wavelength λ / (m)	Frequency ν / (Hz)	Energy E / (J mol^{-1})	Transitions caused	Information gained
Ultraviolet	$10^{-8} - 10^{-10}$	$10^{15} - 10^{17}$	$10^{5} - 10^{7}$	Ionisation of atoms and molecules	Electronic structure and bond energies of molecules
Visible	10^{-7}	10^{14}	10^{5}	Transitions of outer shell electrons of an atom or molecule	Electronic structure of atoms and molecules
Violet	420 nm	7.1×10^{14}			
Blue	470 nm	6.4×10^{14}			
Green	530 nm	5.7×10^{14}			
Yellow	580 nm	5.2×10^{14}			
Orange	620 nm	4.8×10^{14}			
Red	700 nm	4.3×10^{14}			
Infrared	$10^{-7} - 10^{-4}$	$10^{12} - 10^{14}$	$10^{3} - 10^{5}$	Transitions in vibrational and rotational energy states of a molecule	Internuclear distance, force constants of bonds
Microwave	$10^{-4} - 10^{-2}$	$10^{10} - 10^{12}$	$10 - 10^{3}$	Transitions in rotational energy states of a molecule	Internuclear distances
Radio-frequency	$10^{-2} - 10$	$10^{7} - 10^{10}$	$10^{3} - 10$	Reversal of spin of an electron	Electronic structure of paramagnetic substances
				Reversal of spin of a nucleus	Structure of organic compounds

12.3 UV-Visible Spectroscopy

Any experiment in spectroscopy involves either the absorption or the emission of photons from the species under study. The outcome of the experiment is a record of the frequencies absorbed or emitted by the species, and also the extent to which the absorption took place at each frequency. This record is called a *spectrum*. There are two ways to obtain a spectrum. The first is to make light incident on a sample of the species and see what frequencies are absorbed. The resulting spectrum is called an *absorption spectrum* and the technique is called *absorption spectroscopy.* The second method is to excite the species by supplying it energy, and measure the frequencies given out. In this case, the result is an *emission spectrum* and the technique is called *emission spectroscopy.* In this chapter we will deal only with absorption spectroscopy.

12.3.1 Instrumentation

Figure 12.3 shows a schematic diagram for the instrumentation required to carry out such an experiment. A light source that emits radiation in the ultraviolet and visible range is selected. The emitted radiation enters a wavelength selector, which allows only certain wavelengths of light to pass through it. This light is split up into parts: one part passes through a tube containing the absorbing sample, and the part through a tube containing a reference. The reference is generally the solvent in which the sample was made up. A photodetector compares the light emitted by both the sample and the reference in order to eliminate any interference effects due to the solvent. The light is converted into an electric signal and displayed on a readout device.

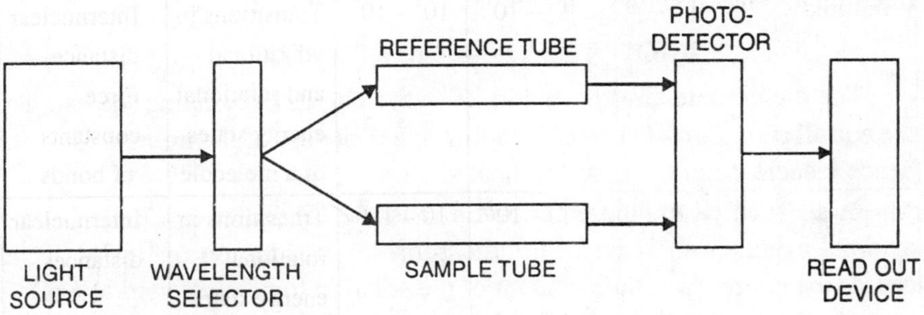

Figure 12.3: Schematic diagram of the instrumentation required for absorption spectroscopy

12.3.2 The Beer-Lambert law

Absorption spectroscopy can be used to quantify the absorbing species present in the sample. Simple reasoning suggests that the greater the quantity of absorbing species present, the greater will be the extent to which the incident light will be absorbed. This logic has been captured in a mathematical equation known as the *Beer-Lambert Law*:

$$A = \varepsilon b C \tag{12.5}$$

where A is the absorbance of the sample

ε is a constant called molar absorptivity that is characteristic of the absorbing species

b is the distance that the incident radiation travels through the sample

C is the concentration of the absorbing species in the sample.

Absorbance has no units. If b is expressed in cm and C in mol dm^{-3} then ε has units of mol^{-1} dm^{3} cm^{-1}.

The Beer Lambert Law is a fundamental law of spectroscopy. Such is its importance that it is often called the 'ABC of spectroscopy', a reference to equation (12.5) which was earlier written as $A = a\, b\, C$.

The Beer-Lambert Law is valid only under certain conditions.

- The light incident on the sample must be monochromatic, i.e., it must consist of one wavelength only. However, it often happens that the wavelength selector allows a band of radiation to pass through it. This band consists of a set of wavelengths very close to each other. If the absorbance of the sample varies widely with each of these wavelengths then a deviation is seen. A plot of absorbance versus concentration will not be linear.
- The concentration of the absorbing species must be 10^{-3} M or less.
- The absorbing species must be in stable chemical equilibrium in the sample.

Consider the equilibrium established between chromate ions and dichromate ions in acidic solution.

$$2CrO_4^{2-} + 2H^+ \longleftrightarrow Cr_2O_7^{2-} + H_2O$$

The dichromate absorbs at 450 nm wavelength. On diluting the sample, the equilibrium shifts to the left, leading to a depletion of dichromate ions and hence a decrease in the absorbance at 450 nm.

In all such cases, the equilibrium must be controlled by adding some external agent. In the case of the dichromate ion, addition of 0.05M KOH causes the entire chromium content of the solution to be converted to CrO_4^{2-} ions, thus preventing the establishment of any equilibrium.

If an absorbing species is present in an acid-base equilibrium then the pH and ionic strength must be controlled. The pH should be at least three units more or less than the pK_a value of the monobasic acid. When the absorbing species is a complex ion, the concentration of the free ligand must be at least a hundred times more than that of the metal ion.

12.4 Acid-base Indicators

An acid-base indicator is a compound that has a certain colour in acidic medium and another colour in basic medium. This property is made use of in acid-base titrations to determine the end point. An acid-base indicator is generally a water-soluble, weakly acidic organic molecule. In aqueous solution, it exists in the following equilibrium:

$$\underset{}{H_2O(aq)} + \underset{\text{acid}}{HIn(aq)} \longleftrightarrow H_3O+(aq) + \underset{\text{conjugate base}}{In\text{-}(aq)} \qquad (12.6)$$

The equilibrium constant is given by

$$K_{eq} = \frac{[H_3O+][In-]}{[HIn][H_2O]} \tag{12.7}$$

If the solution is very dilute, the activity of water will be nearly unity. Equation (12.7) then defines the indicator dissociation constant, K_{In}.

$$K_{In} = \frac{[H_3O+][In-]}{[HIn]} \tag{12.8}$$

Rearranging equation (12.8), we get

$$pH = pK_{In} + \log \frac{[In-]}{[HIn]} \tag{12.9}$$

Absorption spectroscopy can be applied to obtain the value of pK_{In}. When the pH is less than pK_{In} the indicator is mainly in the acidic form and the species HIn is responsible for the absorbance reading, A_{HIn}. When the pH is more than pK_{In}, the indicator is mainly in the basic form and the species In_{-} is responsible for the absorbance reading, A_{In-}. At some intermediate pH value, the absorbance, A, is the sum of the individual absorbances due to HIn and In. If α is the degree of dissociation of the indicator then

$$\frac{\alpha}{1-\alpha} = \frac{[In-]}{[HIn]} \tag{12.10}$$

and

$$A = (1-\alpha)A_{HIn} + \alpha A_{In-} \tag{12.11}$$

Rearranging equation (12.11), we get

$$\frac{\alpha}{1-\alpha} = \frac{A_{HIn} - A}{A - A_{In2}} \tag{12.12}$$

Substituting equation (12.12) in equations (12.10) and (12.9), we get

$$pH = pK_{In} + \log \frac{A_{HIn} - A}{A - A_{Inz}} \tag{12.13}$$

Thus a plot of log $\{(A_{HIn} - A)/(A - A_{In-})\}$ versus pH will be a straight line with pK_{In} as the intercept on the abscissa.

12.5 Spectroscopic Study of Complex Formation

Complex formation between a metal ion and a ligand can be studied using spectroscopy if at least one out of the three – the metal ion, the ligand or the complex – absorbs strongly at a particular wavelength. The molar ratio of metal ion to ligand in the complex can be determined either by *photometric titration* or by *Job's method of continuous variations*.

12.5.1 Photometric titration

The concentration of the metal ions is kept fixed and the concentration of the ligand is incremented stepwise. A plot of absorbance versus moles of ligand added will consist of two straight lines that intersect at the end point. The possible shapes of the titration curves are shown in Figure 12.4. Curve 12.4(a) is obtained when the ligand is the only absorbing species. Curve 12.4(b) is obtained when only the complex ion absorbs. Curve 12.4(c) results when only the metal ion absorbs. When both the metal ion and the ligand absorb, curve 12.4(d) is obtained.

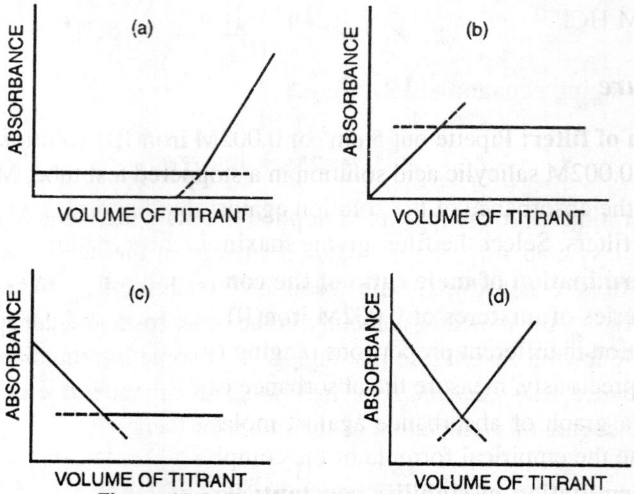

Figure 12.4: Photometric titration curves.

12.5.2 Job's method of continuous variation

In this method, the sum of the molar concentrations of the metal ion and the ligand are kept constant as their ratio is varied. A bell-shaped curve is obtained on plotting absorbance versus mole ratio. The peak on the curve corresponds to the mole ratio in the complex.

EXPERIMENTS

Experiment 12.1

Aim: Determination of the composition and stability constant of a complex formed between iron (III) ions and salicylic acid by Job's method of continuous variation.

Theory

Phenolic compounds like salicylic acid from coloured complexes with ferric salts. When equimolar amounts of the two solutions are mixed in varying proportions, the maximum amount of complex is formed in equilibrium. When the proportions of the reactants correspond to the empirical formula of the complex, absorption is maximum. The optimum pH range for the stability of the iron(III)-salicylic acid complex is 2.6 – 2.8. This is obtained by using solutions made in 0.002M HCl.

Requirements

Colorimeter, 0.002M HCl, 0.002M iron(III) ammonium sulphate in 0.002M HCl, 0.002M salicylic acid in 0.002M HCl, saturated solution of salicylic acid in 0.002M HCl.

Procedure

Selection of filter: Pipette out 5 cm^3 of 0.002M iron(III) solution along with 5 cm^3 of 0.002M salicylic acid solution in a stoppered test tube. Mix well and measure the absorbance of the solution against distilled water as blank using different filters. Select the filter giving maximum absorbance.

Determination of mole ratio of the complex: Prepare in stoppered test tubes a series of mixtures of 0.002M iron(III) solution and 0.002M salicylic acid solution in different proportions ranging from 1:9 to 9:1. Using the filter selected preciously, measure the absorbance of each mixture.

Plot a graph of absorbance against molar composition of the mixtures. Determine the empirical formula of the complex.

Determination of stability constant: Pipette out 1.0, 2.0, 3.0, 4.0 and 5.0 cm^3 of 0.002M iron(III) solution in five different stoppered test tubes. Add 5.0 cm^3 of saturated salicylic acid to each test tube. Make up the volume to 10 cm^3 with 0.002M HCl in case of the first four solutions. Measure the absorbance of each solution using the filter selected preciously.

Plot a graph of absorbance against the concentration of iron (III) solution and determine the molar absorptivity of the complex.

From the various mixtures prepared for the determination of mole ratio of the complex, select the mixture having maximum absorbance. Calculate the concentration of the complex present in the solution. Calculate the concentration of uncombined salicylic acid and iron (III) in this mixture, and hence the stability constant of the complex.

Observations and calculations

Selection of λ_{max}

Wavelength λ (nm)	Absorbance
420	
470	
535	
620	
670	

Therefore, λ_{max} = ______

Determination of mole ratio of the complex

Obs. No.	Volume of 0.002M Fe(III) in 0.002M HCl	0.002M salicylic acid in 0.002M HCl	(cm^3) 0.002M HCl	Absorbance at λ = ______
1	0.5	4.5	5.0	
2	1.0	4.0	5.0	
3	1.5	3.5	5.0	
4	2.0	3.0	5.0	
5	2.5	2.5	5.0	
6	3.0	2.0	5.0	
7	3.5	1.5	5.0	
8	4.0	1.0	5.0	
9	4.5	0.5	5.0	

Maximum absorbance A = ______

Ratio of Fe(III): salicylic acid in this solution = __:__

Therefore, empirical formula of the complex = __:__

Original concentration of Fe (III) in this solution = **B** M

Determination of stability constant

Conc. of Fe(III) (M)	Volume of 0.002M Fe(III) in 0.002M HCl	Saturated salicylic acid in 0.002M HCl	(cm^3) 0.002M HCl	Absorbance at λ = ______
	1.0	5.0	4.0	
	2.0	5.0	3.0	
	3.0	5.0	2.0	
	4.0	5.0	1.0	
	5.0	5.0	0.0	

From the graph of absorbance against the concentration of iron(III) solution,

Slope = molar absorptivity ε = ______

Concentration of Fe(III) in complex $[\text{Fe(III)}]_{\text{complex}}$

$= A/\varepsilon = C =$ ______M

Concentration of Fe(III) left unreacted $[\text{Fe(III)}]_{\text{unreacted}} = (B - A)$ M

Since the complex is formed in 1:1 ratio of Fe(III) and salicylic acid,

Concentration of salicylic acid left unreacted $[\text{SA}]_{\text{unreacted}}$

= Concentration of Fe(III) left unreacted

$= (B - A)$ M

Stability constant $K = [\text{Fe(III)}]_{\text{complex}}/[\text{Fe(III)}]_{\text{unreacted}} \cdot [\text{SA}]_{\text{unreacted}}$

Result

Stability constant of the iron(III)-salicylic acid complex = ______

Experiment 12.2

Aim: To determine the ionisation constant of bromophenol blue.

Requirements

Spectrophotometer, solid bromophenol blue, 1M solutions of CH_3COONa, HCl and NaOH.

Procedure

Prepare the stock indicator solution. Moisten 0.01g of powdered bromophenol blue with 2-3 drops of ethanol, dissolve completely in 0.15 cm^3 of 1M NaOH solution and make up to 100 cm^3 with distilled water.

Standardise the pH meter.

[A] Absorption spectra of acidic and basic forms of bromophenol blue

Make up a solution A containing 20 cm^3 HCl, 5 cm^3 indicator and 25 cm^3 distilled water. Make another solution B with 20 cm^3 NaOH, 5 cm^3 indicator and 25 cm^3 distilled water. Dilute the solutions to 100 cm^3.

Measure the pH values of each solution using a recording spectrophotometer. Record the spectra of these solutions (against distilled water blank) in the range 400-650 nm. For each solution, determine λ_{max} and absorbance at λ_{max}.

[B] Prepare the following solutions in six 100 cm^3 volumetric flasks and measure the pH of each and the absorbance (at the two λ_{max} values obtained in [A]) of each solution.

Flask	*Volume of*			
no.	*CH_3COONa*	*HCl*	*Indicator*	*Distilled water*
1	20.0	8.0	10.0	
2	20.0	10.0	10.0	
3	20.0	12.0	10.0	to make up
4	20.0	14.0	10.0	to the mark
5	20.0	16.0	10.0	
6	20.0	18.0	10.0	

[C] Plot the absorbance of each solution against the pH for each of the two wavelengths. Connect the points with a smooth curve. The mid-point of each curve corresponds to equal concentration of the acid and the base forms of the indicator.

Evaluate the expression log $(A_{HIn} - A)/(A - A_{In^-})$ at each λ_{max} and plot against the pH of the solution. From the intercept, calculate the pK_a of the indicator.

Observations and Calculations

Wavelength	Absorbance	
λ (nm)	Solution *A*	Solution *B*
400 to 650		

Solutions	Absorbance at		pH
	λ_{max} = 440 nm	λ_{max} = 590 nm	
1			
2			
3			
4			
5			
6			
A (A_{HIn})			
B (A_{In^-})			

At λ_{max} = 440 nm, A_{HIn} = _____ and A_{In^-} = _____

Solns	$A_{HIn} - A$	$A - A_{In^-}$	$(A_{HIn} - A)/(A - A_{In^-})$	$\log (A_{HIn} - A)/(A - A_{In^-})$	$pK_a = \text{pH} - \log (A_{HIn} - A)/(A - A_{In^-})$
1					
2					
3					
4					
5					
6					

Mean pK_a = _____

Do the same for λ_{max} = 590 nm.

From calculations,

pK_a at λ_{max} = 440 nm is _____

pK_a at λ_{max} = 590 nm is _____

Mean pK_a = _____

Therefore, K_a = antilog($-pK_a$) = _____

From the graph of $\log (A_{HIn} - A)/(A - A_{In^-})$ versus pH.

pK_a at λ_{max} = 440 nm is (x- intercept) = _____

pK_a at λ_{max} = 90 nm is (x- intercept) = _____

Mean pK_a = _____

Therefore, K_a = antilog($-pK_a$) = _____

Result

The ionisation constant for bromophenol blue is K_a = _____ by calculations and _____ by graph.

Experiment 12.3

Aim: To determine the dissociation constant of an indicator, methyl red, by spectrophotometric method.

Requirements

Spectrophotometer, pH meter with glass electrode assembly, 0.04M CH_3COONa, 0.02M CH_3COOH, 0.1M HCl, 95% ethanol.

Procedure

Prepare a stock solution of methyl red by dissolving 0.1g of pure crystalline methyl red in 30 cm^3 of 95% ethanol and making up to 100 cm^3 with distilled water. Then prepare the following solutions:

Stock solution of indicator 5 cm^3 of stock solution of indicator made up to 100 cm^3 with distilled water.

[A] Solution A 10 cm^3 of stock solution of the indicator , 10 cm^3 of 0.1M HCl diluted to 100 cm^3 with distilled water. (pH of this solution is about 2 and the indicator is entirely in the molecular form HA).

Measure out 40.0 cm^3, 25.0 cm^3 and 10.0 cm^3 of solution *A* into separate 50 cm^3 volumetric flasks and dilute in each case to the mark with 0.01M HCl.

[B]Solution B 10 cm^3 of solution *A*, 25 cm^3 of 0.04M CH_3COONa made up to 100 cm^3 with distilled water. (The pH of this solution is about 8 and the indicator is entirely in the ionic form *A*-).

Measure out 40.0 cm^3, 25.0 cm^3 and 10.0 cm^3 of solution B into separate 50 cm^3 volumetric flasks and dilute in each case to the mark with 0.01M CH_3COONa.

[C] Now prepare the following buffered solutions containing the same quantity of indicator:

Solutions	Volume of (cm^3)			
	0.04M CH_3COONa	0.02M CH_3COOH	Indicator	Distilled water
1	25.0	50.0	10.0	
2	25.0	25.0	10.0	to make up
3	25.0	10.0	10.0	to the mark
4	25.0	5.0	10.0	

Determine the pH values of each of these four solutions with pH meter.

Determine the absorbance of each of the buffer solutions and of solutions *A* and *B* against a blank of distilled water over the range 400-600 nm at intervals of 10 nm. Plot absorbance against wavelength and determine the wavelength of maximum absorption for each solution. All these curves intersect at a common point known as the isobestic point. At this wavelength, the observed absorbance is independent of relative amounts of molecular (HA) and ionic (A^-) forms present but depends only on the total amount of indicator present in the solution.

Plot absorbance versus relative concentration of the indicator for each case in **[A]**, **[B]** and **[C]** and obtain the values of the absorbances at relative concentration of 1.0.

Calculate the relative amounts of the acid and base forms of the indicator in the solutions in **[C]** as a function of pH and hence calculate the pK_a value of the indicator.

Observations and Calculations

Wavelength	Absorbance	
λ (nm)	Solution *A*	Solution *B*
400 to 600		

Solns	Volume of (cm^3)			Absorbance		pH
	0.04M CH_3COONa	0.02M CH_3COOH	Indicator	A_A λ =520 nm	A_B λ=430 nm	
1	25.0	50.0	10.0			
2	25.0	25.0	10.0			
3	25.0	10.0	10.0			
4	25.0	5.0	10.0			

Solutions	Relative concentration	Absorbance	
		λ=520 nm	λ=430 nm
A_1	0.8		
A_2	0.5		
A_3	0.2		
B_1	0.8		
B_2	0.5		
B_3	0.2		

From the graph of absorbance versus relative concentration for each of the two sets of solutions *A* (A_1 A_2 A_3) and *B* (B_1 B_2 B_3) at the wavelengths 520nm and 430 nm,

$d_{A(HMR)}$ = slope of absorbance of *A* versus relative concentration at λ = 520 nm = ______

$d_{A(MR^-)}$ = slope of absorbance of *A* versus relative concentration at λ = 430 nm = ______

$d_{B(HMR)}$ = slope of absorbance of *B* versus relative concentration at λ = 520 nm = ______

$d_{B(MR-)}$ = slope of absorbance of B versus relative concentration at $\lambda = 430$ nm = ______

where $d_{A(HMR)}$, $d_{A(MR-)}$, $d_{B(HMR)}$, $d_{B(MR-)}$ are absorbance values at relative concentration 1.0.

The relative amounts of HMR and MR– can be calculated from the two equations

$$A_A = d_{A(HMR)} [HMR] + d_{A(MR-)} [MR-]$$
$$A_B = d_{B(HMR)} [HMR] + d_{B(MR-)} [MR-]$$

Solving these equations simultaneously, we get

$$[HMR] = \frac{d_{B(MR)} A_A - d_{A(MR)} A_B}{d_{B(MR)} d_{A(HMR)} - d_{A(MR-)} d_{B(HMR)}}$$

$$[MR-] = \frac{d_{B(HMR)} A_A - d_{A(HMR)} A_B}{d_{B(HMR)} d_{A(MR-)} - d_{A(HMR)} d_{B(MR-)}}$$

Solns	[HMR]	[MR–]	[MR–]/[HMR]	log [MR-]/[HMR]	pH	pK_a = pH – log [MR-]/[HMR]
1						
2						
3						
4						
5						
6						

Mean pK_a = ______

K_a = antilog $(-pK_a)$ = ______

Result

Dissociation constant of methyl red indicator K_a = ______

Experiment 12.4

Aim: To determine the stability constant of the complex formed between Fe(III) ions and 5-sulphosalicylic acid at pH 2 and pH 3 by colorimetric method.

Requirements

Colorimeter, 0.001 M Fe^{3+} solution (ferric ammonium sulphate) prepared in 0.01 N HCl, 0.001 M Fe^{3+} solution (ferric ammonium sulphate) prepared in 0.001 N HCl, 0.001 M 5-sulphosalicylic acid prepared in 0.01 N HCl, 0.001 M 5-sulphosalicylic acid prepared in 0.001 N HCl.

Theory

5-sulphosalicylic acid forms a 1:1 complex with iron (III) ions. Using Job's method of continuous variation, the concentration of the metal ions and 5-sulphosalicylic acid required for complex formation can be determined and hence the stability constant of the complex can be calculated. The free energy of the reaction can be calculated from the thermodynamic formula relating equilibrium constant to the free energy.

$$\Delta G = -RT \ln K$$

Procedure

For pH 2

Prepare a set of solutions as given below using 0.001 M Fe^{3+} solution and 0.001 M 5-sulphosalicylic acid prepared in 0.01 N HCl.

Volume of (cm^3)		Absorbance at λ_{max} = ____ nm
0.001 M Fe^{3+}	0.001 M 5-SSA	
1	9	
2	8	
3	7	
4	6	
5	5	
6	4	
7	3	
8	2	
9	1	

Using the (5 cm^3 Fe^{3+} + 5 cm^3 5-SSA) solution, select the λ_{max}. Record the absorbance of each solution at the λ_{max} selected. Plot a graph of absorbance versus volume of Fe^{3+}. Calculate the stability constant of the complex.

For pH 3

Prepare a set of solutions as given below using 0.001 M Fe^{3+} solution and 0.001 M 5-sulphosalicylic acid prepared in 0.001 N HCl.

Volume of (cm^3)		Absorbance at λ_{max} = ____ nm
0.001 M Fe^{3+}	0.001 M 5-SSA	
1	9	
2	8	
3	7	
4	6	
5	5	
6	4	
7	3	
8	2	
9	1	

Record the absorbance of each solution at the λ_{max} selected. Plot a graph of absorbance versus volume of Fe^{3+}. Calculate the stability constant of the complex.

Observations and calculations

Determination of λ_{max}

Wavelength (nm)	Absorbance
660	
620	
535	
470	
420	

λ_{max} = ____ nm

For pH 2

Volume of (cm^3)		Absorbance at λ_{max} = ____ nm
0.001 M Fe^{3+}	0.001 M 5-SSA	
1	9	
...	...	
9	1	

Maximum absorbance A = _____

Ratio of Fe(III):5-sulphosalicylic acid in this solution = __ : __

Therefore, empirical formula of the complex = __ : __

Original concentration of Fe (III) in this solution = _______ M

Determination of stability constant

It can be assumed that the complex is almost completely formed at 2 : 8 ratio of Fe(III) : 5-sulphosalicylic acid since the concentration of the ligand is far in excess of that required for complex formation.

Molar absorptivity ε = absorbance/concentration = _______ $mol^{-1}\ dm^3\ cm^{-1}$

Concentration of Fe(III) in complex = $[Fe(III)]_{complex}$

= absorbance/molar absorptivity = _______ M

Concentration of Fe(III) left unreacted = $[Fe(III)]_{unreacted}$ = _____ M

Since the complex is formed in 1:1 ratio of Fe(III) and 5-sulphosalicylic acid,

Concentration of 5-sulphosalicylic acid left unreacted $[5\text{-}SSA]_{unreacted}$ = $[Fe(III)]_{unreacted}$ = ____ M

$$\text{Stability constant } K = \frac{[Fe(III)]_{complex}}{[Fe(III)]_{unreacted}\ [5-SSA]_{unreacted}}$$

= _____

For pH 3

The calculations for pH 3 can be done in a fashion similar to that for pH 2.

Results

1. Empirical formula of the complex formed between Fe(III) and 5-sulphosalicylic acid = __ : __
2. Stability constant of the complex at pH 2 = ______
3. Stability constant of the complex at pH 3 = ______

Experiment 12.5

Aim: To determine the stability constant of the complex formed between Cu(II) ions and 5-sulphosalicylic acid between pH 3-5 by spectrophotometric method and hence calculate the free energy of formation of the complex.

Requirements

0.2 M $CuSO_4$, 0.2 M 5-sulphosalicylic acid, 1 N NaOH, 2 M Na_2SO_4, 0.1 M H_2SO_4, spectrophotometer, pH meter, buffer solutions of pH 4.01 and 9.2.

Theory

5-sulphosalicylic acid forms a leaf-green complex with copper (II) ions in the pH range of 3 to 5 of the type CuR. Using Job's method of continuous variation, the concentration of the metal ions and 5-sulphosalicylic acid required for complex formation can be determined and hence the equilibrium constant of the reaction can be calculated. The free energy of the reaction can be calculated from the thermodynamic formula relating equilibrium constant to the free energy.

$$\Delta G = -RT \ln K$$

Procedure

Set-up and standardise the pH meter using the buffer solutions.

Prepare solution *A* as given below.

Solution A: Pipette out 10 cm^3 of 2 M Na_2SO_4 solution, 2 cm^3 of copper sulphate solution and 5 cm^3 of 5-sulphosalicylic acid solution in a 25 cm^3 beaker. Dip the glass electrode of the pH meter in the solution and add 1 N NaOH drop wise until the pH rises to 5. If the pH exceeds 5, adjust it by adding a few drops of 0.1 M H_2SO_4. Transfer the solution quantitatively into a 50 cm^3 standard volumetric flask and make up to the mark with distilled water.

Record the absorption spectrum of solution *A* from 600 nm to 800 nm at 10 nm intervals using a spectrophotometer. Plot a graph of absorbance of solution *A* versus wavelength. Select λ_{max} for solution *A* where λ_{max} is the wavelength at which maximum absorbance occurs.

Prepare a set of solutions as given below.

Pipette out in a 25 cm^3 beaker the volumes of sodium sulphate solution, copper sulphate solution and 5-sulphosalicylic acid solution according to the table given below. Adjust the pH to 5 by adding 1 N NaOH drop wise. Transfer the solution quantitatively into a 50 cm^3 standard volumetric flask and make up to the mark with distilled water. Record the absorbance of each solution at the λ_{max} selected for solution *A* against its corresponding blank solution. The blank for each solution should contain the same volume of copper sulphate solution diluted to 50 cm^3. For example, the blank for the first solution given below is 1 cm^3 of 0.2 M $CuSO_4$ diluted to 50 cm^3.

Volume of (cm^3)			Absorbance at λ_{max} = ___ nm
0.2 M $CuSO_4$	0.2 M 5-SSA	2 M Na_2SO_4	
1	9	10	
2	8	10	
3	7	10	
4	6	10	
5	5	10	
6	4	10	
7	3	10	
8	2	10	
9	1	10	

Plot a graph of absorbance of the above solutions versus volume of Cu (II). Find the composition of the mixture for which maximum absorbance takes place and hence determine the empirical formula of the complex.

Calculate the stability constant of the complex and hence its free energy of formation.

Observations and calculations

Absorbance values of solution *A* from 600 nm to 800 nm

Wavelength (nm)	Absorbance of Soln *A*
600	
...	
800	

λ_{max} for Solution *A* = ____ nm

Maximum absorbance *A* = _____

Ratio of Cu(II):5-sulphosalicylic acid in this solution = __ : __

Therefore, empirical formula of the complex = __ : __

Original concentration of Cu (II) in this solution = ______ M

Determination of stability constant

It can be assumed that the complex is almost completely formed at 2 : 8 ratio of Cu(II) : 5-sulphosalicylic acid since the ligand concentration is far in excess of that required for complex formation.

Molar absorptivity ε = absorbance/concentration = ______ mol^{-1} dm^3 cm^{-1}

Concentration of Cu(II) in complex = $[Cu(II)]_{complex}$

= absorbance/molar absorptivity

= __________M

Concentration of Cu(II) left unreacted = $[Cu(II)]_{unreacted}$ = _____ M

Since the complex is formed in 1:1 ratio of Cu(II) and 5-sulphosalicylic acid,

Concentration of 5-sulphosalicylic acid left unreacted $[5\text{-SSA}]_{unreacted}$ = $[Cu(II)]_{unreacted}$ = _______M

$$\text{Stability constant } K = \frac{[Cu(II)]_{complex}}{[Cu(II)]_{unreacted}\ [5-SSA]_{unreacted}}$$

= __________

Free energy of formation of the complex $\Delta G = -RT \ln K$

$\Delta G = -RT \ln K$ = ____ kJ mol^{-1}

Results

1. Empirical formula of the complex formed between Cu(II) and 5-sulphosalicylic acid = __ : __
2. Stability constant of the complex = ______
3. Free energy of formation of the complex = ____ kJ mol^{-1}

Experiment 12.5

Aim: To study the complex formation between Ni(II) ions and 1,10-phenanthroline.

Theory

Nickel ions can form three complex ions with 1,10-phenanthroline as follows:

$$Ni^{2+} + n\ (1,10\text{-phenanthroline}) \longleftrightarrow [Ni\ (Ph)_n]^{2+}$$

where n can be 1, 2 or 3.

Requirements

A spectrophotometer in the visible range, 0.1 M nickel sulphate, solid 1,10-phenanthroline.

Procedure

Prepare the following solutions in graduated test tubes:

Volume of 0.1 M $NiSO_4$ soln / (cm^3)	Volume of 0.1 M 1,10-phenanthroline soln/ (cm^3)
1.0	9.0
2.0	8.0
2.5	7.5
3.0	7.0
3.5	6.5
4.0	6.0
5.0	5.0
6.0	4.0
7.0	3.0
8.0	2.0
9.0	1.0

Since 1,10-phenanthroline is sparingly soluble in water, it is advisable to weigh the quantity of 1,10-phenanthroline equivalent to the required volume of 0.1 M solution, add the required volume of 0.1 M $NiSO_4$ solution and make the solution to 10 cm^3 with distilled water. For example, to make the first solution given in the above table, weigh 0.1784 g of 1,10-phenanthroline, add 1.0 cm^3 of 0.1 M $NiSO_4$ and 9.0 cm^3 of distilled water. Measure the absorbance, *A*, of each solution at 528 nm, 580 nm and 620 nm against a blank of distilled water.

At each wavelength, plot the graph of the absorbance versus the volume of nickel sulphate solution.

Dilute 1.0, 2.0, 4.0, 6.0 and 8.0 cm^3 of 0.1 M $NiSO_4$ solution to 10 cm^3 with distilled water. Measure the absorbance, *A*′, of each solution at 528 nm, 580 nm and 620 nm against a blank of distilled water. At each wavelength, plot the graph of the absorbance versus the volume of nickel sulphate solution.

Plot (*A* – *A*′) versus the volume of nickel sulphate solution and from the curves obtained, determine the formulae of the complex ions.

Result

Three complexes are formed between nickel (II) ions and 1,10-phenanthroline in the mole ratios of ____ , ___ and ____ which have λ_{max} at ___ nm, ___ nm and ___ nm respectively.

Appendix

Review Questions

1. How is the choice of an indicator made?
2. What is meant by indicator range?
3. What is the universal indicator?

Table 12.1: Acid-base indicators

Indicator	Approximate pH range	Colour change	pK_{In}	Preparation
Crystal violet	0.0 – 1.8	Yellow – blue	–	0.02% in water
Malachite green	0.2 – 1.8	Yellow – blue	–	0.1% in water
Thymol blue	1.2 – 2.8	Red – yellow	1.7	0.1 g in 21.5 ml 0.01N
	8.0 – 9.6	Yellow - blue	8.9	NaOH + 299.5 ml water
Metanil yellow	1.2 – 2.4	Red – yellow	–	0.01% in water
Bromophenol blue	3.0 – 4.6	Yellow – blue	4.1	0.1 g in 14.9 ml 0.01N NaOH + 235 ml water
Congo red	2.0 – 5.0	Blue – red	–	0.1% in water
Methyl orange	3.2 – 4.4	Red – yellow	3.7	0.01% in water
Methyl red	4.8 – 6.0	Red – yellow	5.0	0.02 g in 60 ml ethanol + 40 ml water
Bromocresol purple	5.2 – 6.8	Yellow – purple	6.1	0.1 g in 18.5 ml 0.01N NaOH + 231.5 ml water
Alizarin	5.6 – 7.2	Yellow – red	–	0.1% in methanol
Bromothymol blue	6.0 – 7.6	Yellow – blue	7.1	0.1 g in 16 ml 0.01 N NaOH + 234 ml water
Phenolphthalein	8.2 – 10.0	Colourless to pink	9.6	0.05 g in 50 ml ethanol + 50 ml water

Expected Experimental Results

Experiment 12.1

Stability constant of the iron(III)-salicylic acid complex is approximately 10^4.

Experiment 12.2

The pK_{In} for bromophenol blue is 4.1.

Experiment 12.3

The pK_{In} for methyl red is 5.0.

Experiment 12.4

Empirical formula of the complex formed between Cu(II) and 5-sulphosalicylic acid is 1 : 1.

Stability constant of the complex is approximately 10^3.

Free energy of formation of the complex is approximately 20 kJ mol^{-1}.

Experiment 12.5

The mole ratios of nickel (II) and 1,10-phenanthroline are 1:1, 1:2 and 1:3, which have maximum absorbance at 630 nm, 580 nm and 528 nm respectively.

Hints to Review Questions

1. For an acid-base neutralisation titration, the pH change on both sides of the equivalence point is determined. An indicator that changes colour at or near the equivalence point is selected.
2. The human eye is able to detect a change in the colour of the indicator solution in the interval pH = $pK_{In} \pm 1$, i.e., over approximately two pH units. This is known as the indicator range over which it will appear to change from one colour to another.
3. A universal indicator is a mixture of several indicators which changes colour over a large portion of the pH range. They are usually used for determining approximately the pH of a solution. They are available commercially as solutions or as test papers. One such universal indicator is prepared by dissolving 0.1 g phenolphthalein, 0.2g methyl red, 0.3 g methyl yellow, 0.4 g bromothymol blue and 0.5 g thymol blue in 500 cm^3 of absolute ethanol and adding sodium hydroxide till the solution turns yellow. The colour changes occur at pH 2 (red), pH 4 (orange), pH 6 (yellow), pH 8 (green) and pH 10 (blue).

Further Exploration through the Internet

http://www.spectroscopymag.com/
http://www.s-a-s.org/
http://education.llnl.gov/msds/orgchem/spectroscopy.html
http://spec.jpl.nasa.gov/
http://www.chem.ucla.edu/~webspectra/
http://chipo.chem.uic.edu/web1/ocol/spec/IR.htm
http://dir.yahoo.com/Science/Chemistry/Spectroscopy/
http://www-wilson.ucsd.edu/education/pchem/spectroscopy/
http://physics.nist.gov/Pubs/AtSpec/
http://chipo.chem.uic.edu/web1/ocol/spec/MS.htm
http://www.icmm.csic.es/Fagullo/ramw3_e.htm
http://chemistry.about.com/science/chemistry/cs/spectroscopy/

Pierre Curie

Pierre Curie was born in 1859 in Paris, France. His father, a doctor, educated him at home through private tutoring. By the age of 14, Pierre had shown a remarkable talent for mathematics and spatial geometry. At 16, he started studies at the Faculty of Science at the Sorbonne. At 18, he obtained his Licentiateship (equivalent to a master's degree) in Physics. Lack of money forced him to put off work for his doctorate indefinitely. He became a poorly paid laboratory assistant at the Sorbonne.

Along with his elder brother, Jacques, Pierre studied crystallography. They discovered the phenomenon of piezoelectricity and speculated on the fundamental role of symmetry in the laws of physics.

In 1888, Pierre was appointed supervisor at the School of Physics and Industrial Chemistry at Paris. He was able to resume his own research and turned his interest to the field of magnetism. He aimed to determine whether there were any transitions between the three types of magnetism (dia, para and ferro) with change in temperature. The transition point between paramagnetism and ferromagnetism at a certain temperature is now called the Curie temperature. He discovered that the magnetic coefficients of attraction of paramagnetic bodies vary inversely in proportion to the absolute temperature, a phenomenon now known as Curie's Law. To measure the magnetic coefficients, he constructed several delicate balances. A simplified version of such a balance is still used today and is called the Curie-Cheveneau balance.

In the spring of 1894 Pierre met Marie Sklodowska, a Polish student. On her urging, Pierre wrote up his research on magnetism as a doctoral thesis. In early 1895 Pierre was awarded a doctorate of science.

The marriage of Pierre and Marie in July 1895 marked the beginning of a historic partnership. In 1898 they jointly discovered the elements polonium and radium by the fractionation of pitchblende. After this discovery they decided on a division of labour; Pierre would investigate the properties of radium while Marie would seek better ways to prepare pure compounds. Measuring the action of magnetic fields on the rays that radium gives out, Pierre proposed the existence of positive, negative and neutral particles. Ernest Rutherford later named them alpha, beta and gamma rays, respectively. The Curies' work founded the basis for most of the subsequent research in nuclear physics and chemistry.

Due to Pierre's early unorthodox education he was never accepted completely by the French scientific establishment. On his own part, Pierre despised the politics and flattery required to advance in the small world of Parisian science. In 1902 his application to enter the French Academy of Sciences was rejected. As a result, the Curie's research was done under conditions of much hardship. The laboratory facilities were minimal and the Curies were always under the pressure to teach in order to earn their livelihood.

Vindication came in 1903 when the Curies were awarded the Nobel Prize for Physics jointly with Henri Becquerel. The same year they also won the Royal Society's Davy Medal. The French Academy of Sciences finally opened its doors to Pierre in 1905. His life was cut short abruptly when he was run over by a dray on a rainy April day in Paris in 1906. In 1908, his complete works were published. Marie also wrote a short, eloquent biography on him.

13
Magnetic Properties

13.1 Introduction

As early as 600 BC the Greeks knew that a certain form of iron ore, now known as magnetite or lodestone, had the property of attracting small pieces of iron. The word *lodestone* is derived from an old English word meaning *way* and refers to the fact that a piece of lodestone when suspended freely, always comes to rest pointing in a north-south direction. Chemically, lodestone consists of an iron ore called magnetite whose principal component is the mixed iron oxide Fe_3O_4. The words *magnetite* and *magnetism* are derived from *Magnesia*, the place where magnetic iron ore was first discovered.

13.2 Terminology

Before discussing the magnetic properties of materials, let us spend a little effort understanding the terms that will be employed.

13.2.1 Magnetic Field (H)

If a bar magnet is dipped into iron filings it is noticed that the filings cling in tufts near the ends (Figure 13.1a). Few, if any, filings are attracted to the middle of the bar. The regions near the ends of the magnet where the iron filings are concentrated are called *magnetic* poles. Every magnet has two poles: the north pole (N) and the south pole (S). The magnetic field, *H*, that is exerted by the magnet can be thought of as consisting of magnetic lines of force (Figure 13.1b). They emerge from the north pole of the magnet and re-enter it at the south pole. The SI unit of magnetic field is tesla (T). The CGS unit is gauss.

$$1 \text{ T} = 10^4 \text{ gauss}$$

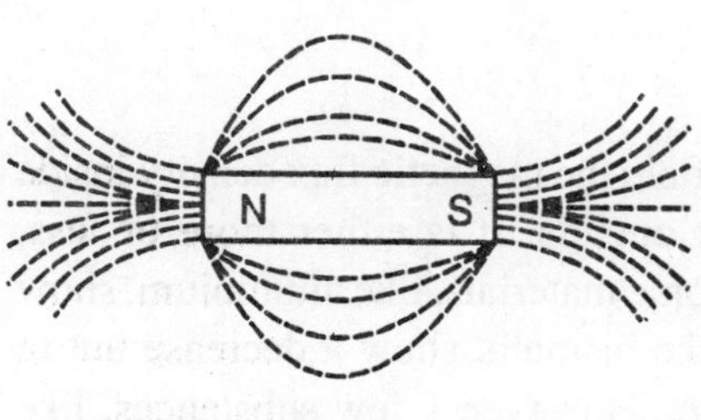

Figure 13.1 (a): Iron filings clustered around a bar magnet

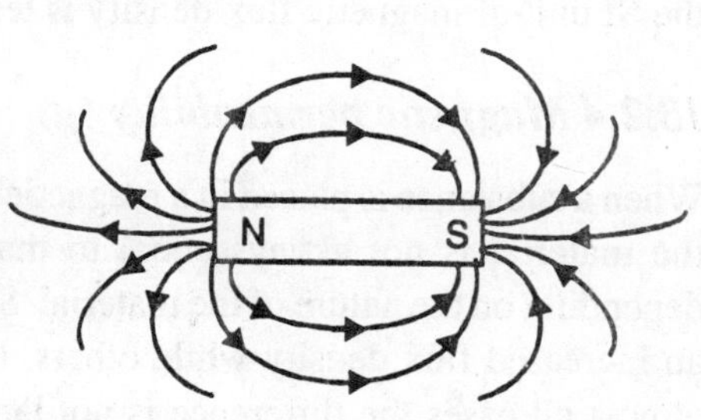

Figure 13.1 (b): Magnetic lines of force

13.2.2 Magnetic moment (μ)

A magnet has two opposite poles at its ends. Hence it behaves like a magnetic dipole. The behaviour of a magnetic dipole in a magnetic field is analogous to the behaviour of an electric dipole in an electric field. When a magnet is placed in magnetic field, *H*, such that it makes an angle θ with the lines of force a torque or turning force, τ, acts on the magnet. The torque tries to align the north-south axis of the magnet with the direction of the magnetic field. This torque depends on the magnetic field as well as on *sin* θ (Figure 13.2)

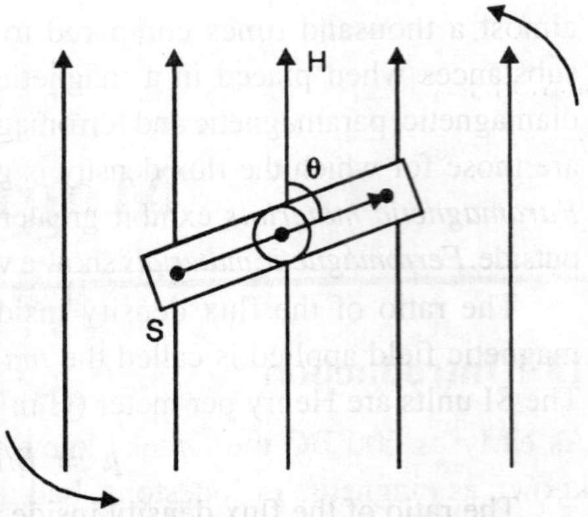

Figure 13.2: Torque acting on a magnet.

$$\tau \propto H \sin\theta$$

$$\tau = \mu H \sin\theta \tag{13.1}$$

where μ is known as the magnetic moment. It is a characteristic property of the material of which the magnet is made.

Since torque and magnetic field are vectors, the magnetic moment is also a vector. It is directed along the axis of the magnet from its south pole to its north pole. The SI unit of magnetic moment is N m T^{-1} which is equivalent to A m^2.

13.2.3 Magnetic flux density (B)

Magnetic flux density is a measure of the number of lines of force that pass through a given area. If an area *A* has its surface perpendicular to a magnetic field of strength *H* then the magnetic flux density passing through that area is given by

$$B = HA \tag{13.2}$$

The flux density at a point can be defined as the flux per unit area when the area is normal to the field about that point. When defined in this manner, the SI unit of magnetic flux density is tesla.

13.2.4 Magnetic permeability (μ)

When a substance is placed in a magnetic field the magnetic flux density inside the material is not always equal to that outside. It is either more or less, depending on the nature of the material. Some materials, like aluminium, show an increased flux density while others, like bismuth, show a decrease but in almost all cases the difference is not large. However, a few substances, like iron, show a dramatic increase in flux density. The flux density inside them is almost a thousand times compared to that outside. This behaviour among substances when placed in a magnetic field leads to their classification as diamagnetic, paramagnetic and ferromagnetic materials. *Diamagnetic materials* are those for which the flux density is greater outside than inside the material. *Paramagnetic materials* exhibit greater flux density inside the material than outside. *Ferromagnetic materials* show a very large flux density inside the material.

The ratio of the flux density inside the material to the strength of the magnetic field applied is called the *magnetic permeability* μ of the material. The SI units are Henry per meter (H m^{-1}).

$$\mu = B/H \tag{13.3}$$

The ratio of the flux density inside the material, B, to that in vacuum, B_0, is called the relative permeability, μ_r, of the material. It is a dimensionless quantity.

$$\mu_r = B/B_0 \tag{13.4}$$

Relative permeability can also be expressed as a ratio of the magnetic permeability of the material, μ, to the magnetic permeability of vacuum, μ_0.

$$\mu_r = \mu/\mu_0 \tag{13.5}$$

It is clear that for diamagnetic materials the relative permeability will be less than unity while for paramagnetic and ferromagnetic materials it will be greater than unity.

13.2.5 Magnetisation (M)

The magnetic flux density inside a material is due to the magnetic dipoles already present in the material as well as those dipoles that are induced when the material is placed in the magnetic field. The induced magnetic dipole moment per unit volume is known as the *magnetisation*, M, of the material.

$$B = \mu_0 (H+M) \tag{13.6}$$

The SI unit of magnetisation is A m^2 m^{-3} or A m^{-1}.

A paramagnetic or ferromagnetic material has greater flux density inside it than outside so its magnetisation is obviously a positive quantity. In contrast, the magnetisation of a diamagnetic material is a negative quantity, which makes its flux density less compared to that outside it.

13.2.6 Magnetic susceptibility (χ)

It has been observed that the stronger the magnetic field, the greater is the magnetisation that it produces.

$$M \propto H$$

$$M = \chi H \tag{13.7}$$

where χ is the proportionality constant. It is known as the *volume magnetic susceptibility* of the material. It is a measure of how easily the material can be magnetised in the presence of a magnetic field. It is a dimensionless quantity.

Substituting equation (13.7) in equation (13.6), we get the magnetic flux density in terms of the volume magnetic susceptibility.

$$B = \mu_0 (H + \chi H)$$

$$B = \mu_0 H (1 + \chi) \qquad (13.8)$$

It is more common to express the magnetic susceptibility of substances in terms of molar magnetic susceptibility, χ_m, which is the product of the volume magnetic susceptibility and the molar volume, V_m.

$$\chi_m = \chi V_m \qquad (13.9)$$

It is clear from equation (13.8) that diamagnetic substances have negative values of magnetic susceptibility while paramagnetic and ferromagnetic substances have positive values. The magnetic susceptibility of ferromagnetic materials is generally of the order of 10^3.

Table 13.1 in Appendix 13 lists the molar magnetic susceptibilities of a few elements, inorganic compounds and organic compounds.

13.3 Modern Views on Magnetism

All matter is ultimately composed of atoms. Atoms consist of electrons orbiting around the nucleus. Besides this, electrons also spin about their own axis. An electron is a charged particle. The movement of a charged particle gives rise to an electric current, which in turn generates a magnetic field. Thus all electrons are tiny magnets that derive their magnetism from their orbital and spin motions.

Diamagnetism

In many substances there are equal numbers of electrons spinning and orbiting in opposite directions so that, in the absence of an external magnetic field, their effects cancel out. When placed in a magnetic field, the electron orbits get disturbed and a tiny magnetic moment arises. This induced magnetic moment is oriented opposite to the applied magnetic field, which causes the flux density inside the material to be less than that outside. Elements and compounds with all paired electrons exhibit diamagnetism, e.g., water, benzene, antimony, etc.

Paramagnetism

Some materials contain atoms or molecules that have unpaired electrons. These atoms or molecules are arranged randomly within the material such that they cancel out each other's tiny magnetic field. When a magnetic field is applied externally, these atoms or molecules align themselves parallel to the field, thus increasing the magnetic flux density inside. When the external magnetic field is removed, the alignment disappears and the old random ordering once again takes over. Transition elements of the Periodic Table (*d*-block and *f*-block elements) contain unpaired electrons in their inner shells and hence exhibit paramagnetism.

Ferromagnetism

Ferromagnetic materials contain atoms that have unpaired electrons, similar to paramagnetic materials. However, these atoms do not orient themselves independently of each other, as is the case with paramagnetic materials. They interact with each other over small distances to spontaneously align their magnetic moments in a certain preferred direction. Groups such as these are called *magnetic domains*. The net magnetic moment of each domain is randomly oriented so that the material as a whole does not have any polarity. This condition is shown in Figure 13.3a.

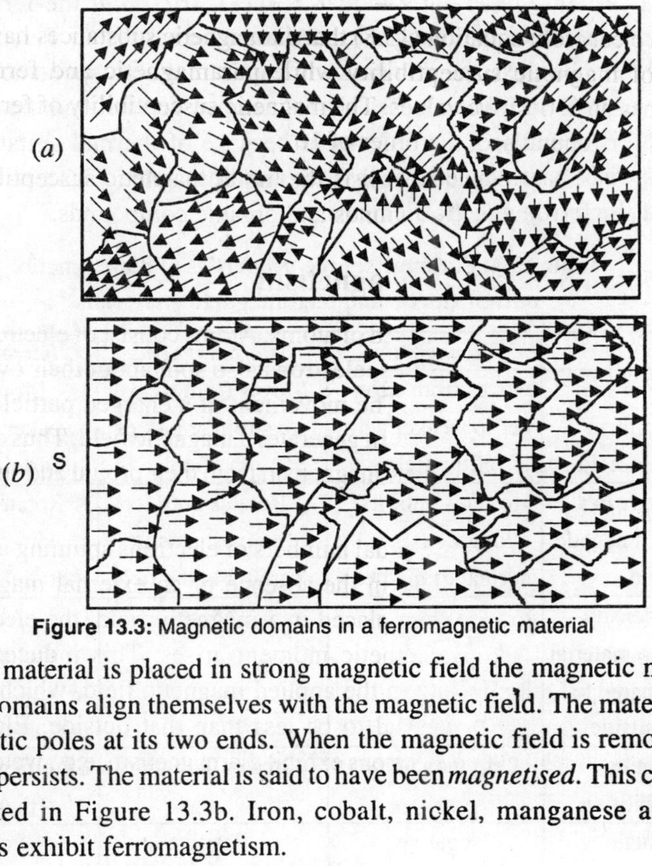

Figure 13.3: Magnetic domains in a ferromagnetic material

If this material is placed in strong magnetic field the magnetic moments of all the domains align themselves with the magnetic field. The material then has magnetic poles at its two ends. When the magnetic field is removed this alignment persists. The material is said to have been *magnetised*. This condition is illustrated in Figure 13.3b. Iron, cobalt, nickel, manganese and their compounds exhibit ferromagnetism.

13.4 Variation of Magnetic Properties with Temperature

Diamagnetism is caused solely due to the presence of paired electrons in the substance. Hence it does not show any variation with temperature.

In 1895 the French physicist, Pierre Curie, proposed an empirical law to show the variation of the magnetic susceptibility of a paramagnetic substance with temperature.

$$\chi = A + \frac{C}{T} \tag{13.11}$$

where A and C are constants and T is the absolute temperature.

Table 13.2: Curie points of the ferromagnetic elements

Element	*Curie point T_C/K*
Fe	1043
Co	1388
Ni	627

Equation (13.11) is known as the *Curie Law* in honour of its discoverer. Above a certain temperature known as the *Curie point* the ferromagnetic behaviour of a substance changes to paramagnetic behaviour. Ferromagnetism arises chiefly because the magnetic moments of atoms spontaneously align themselves to form domains. Raising the temperature beyond the Curie point disrupts this spontaneous arrangement because of thermal agitation. Hence only the weak paramagnetic behaviour remains. Table 13.2 gives the Curie points of the ferromagnetic elements.

Table 13.3: Summarises the properties of diamagnetic, paramagnetic and ferromagnetic materials

Property	*Diamagnetic*	*Paramagnetic*	*Ferromagnetic*
Behaviour in a non-uniform magnetic field	Moves from stronger to weaker part of the field	Moves from weaker to stronger part of the field	Moves from weaker to stronger part of the field
Behaviour in a uniform magnetic field	Rotates until its longest axis is normal to the field	Rotates until its longest axis is parallel to the field	Rotates until its longest axis is parallel to the field
Flux density inside the material as compared to that outside	Less	More	About a thousand-fold more
Magnetic permeability μ	Less than unity	Greater than unity	Significantly greater than unity
Magnetic susceptibility χ	Negative	Positive	Positive
Variation with temperature	Invariant	Decrease with rise in temperature	Has a characteristic temperature called Curie point. If heated beyond the Curie point, the magnetic behaviour is that of a paramagnetic substance.

13.5 Experimental Determination of Magnetic Susceptibility using a Guoy Balance

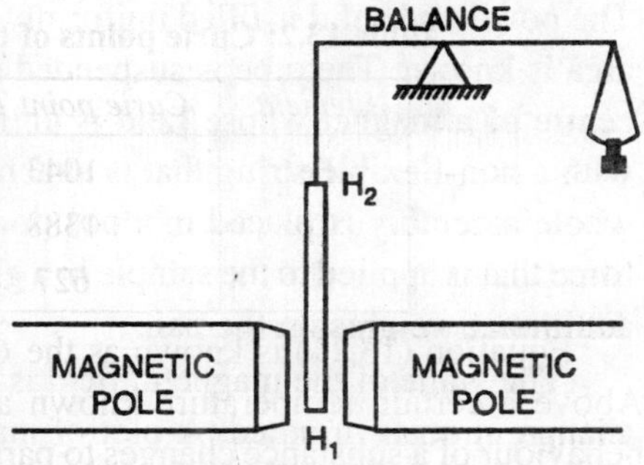

Figure 13.4: The Guoy balance.

The Guoy balance consists of a sensitive mass balance from which hangs a narrow tube (Figure 13.4). The opposite poles of a strong magnet lie on either side of the tube. The sample is powdered and filled into the narrow tube. The tube is weighed. The tube is hung from the balance. If the sample is paramagnetic the tube is pulled down further into the magnetic field. The apparent increase in the mass of the sample is balanced by adding weights to the pan of the Guoy balance. If the substance is diamagnetic, the tube is repelled by the magnetic field and suffers from an apparent decrease in mass. The apparent change in the mass of the sample is proportional to the magnetic susceptibility of the sample.

The magnetic force on the tube along the axis of the tube is given by

$$F = \tfrac{1}{2} \mu_0 \chi S(H_1^2 - H_2^2) \qquad (13.12)$$

where S is the cross-sectional area of the tube

H_1 is the strength of the magnetic field at that end of the tube which is located between the magnet poles

H_2 is the strength of the magnetic field at the other end of the tube

The magnetic force on the tube is balanced by the weight added to the pan of the balance. Hence

$$\text{mg} = \tfrac{1}{2} \mu_0 \chi S(H_1^2 - H_2^2) \qquad (13.13)$$

If the field H_2 is almost negligible then

$$\text{mg} = \tfrac{1}{2} \mu_0 \chi S\, H_1^2 \qquad (13.14)$$

Thus by varying the magnetic field and obtaining a series of corresponding mass values, a plot of mass versus magnetic field can be drawn. The straight line so obtained will have slope equal to $\mu_0 \chi S / 2\text{g}$ from which the value of χ can be calculated.

EXPERIMENTS

Experiment 13.1

Aim: To determine the magnetic susceptibility of a given substance using the Guoy balance.

Requirements

Guoy balance, powdered sample.

Procedure

The powdered solid is filled uniformly into a glass tube whose cross-sectional area is known. The tube is suspended so that the lower end just comes at the centre of a magnet whose field is uniform. The other end of the tube is tied with a non-flexible string that is attached to a pan of a sensitive balance. The whole assembly is placed in a position such that it is free of vibrations. The force that is applied to the sample in a given magnetic field is known by putting additional weights on the pan.

The value of the magnetic field is varied and the corresponding apparent change in mass in noted. A plot of mass m versus magnetic field H applied would give a straight line with slope $\mu_0\chi S/2g$ (equation (13.14)) from which the value of χ can be calculated.

Observations and calculations

To calculate the cross-section area of the tube

Pole gap = ____ cm
Inner diameter of tube d = ____ cm
Inner cross-section area of tube = $\pi (d/2)^2$ = ___ cm^2
Mass of empty tube = ___ g
Mass of tube + powder = ____ g
Mass of powder = ____ g

Current passed through the electromagnet/ (A)	Magnetic field $H/(A\ m^{-1})$	$H^2/(A\ m^{-1})^2$	Mass added to pan m/ (g)

From the graph of m versus magnetic field H, slope = _______

$$\text{Slope} = \mu_0\chi S/2g$$

Therefore, χ = slope × 2g / $\mu_0 S$ = _______

Result

The magnetic susceptibility of the given substance is ______ .

Appendix

Review Questions

1. Explain triplet and singlet states.
2. Tape recorders, cassettes and floppies should not be kept near a television.

Table 13.1: Magnetic Susceptibilities at 298 K

Substance	$\chi / 10^{-6}$	χ_m / (10^{-4} cm^3 mol^{-1})
Water	−90	−16.0
Benzene	−7.2	−6.4
Cyclohexane	−7.9	−8.5
Carbon tetrachloride	−8.9	−8.4
NaCl(s)	−13.9	−3.75
Cu(s)	−96	−6.8
S(s)	−12.9	−2.0
Hg(l)	−28.5	−4.2
$CuSO_4.5H_2O$(s)	+176	+192
$MnSO_4.4H_2O$(s)	+2640	$+2.79 \times 10^3$
$NiSO_4.7H_2O$(s)	+416	+600
$FeSO_4(NH_4)_2SO_4.6H_2O$(s)	+755	$+1.51 \times 10^3$
Al(s)	+22	+2.2
Pt(s)	+262	+22.8
Na(s)	+7.3	+1.7
K(s)	+5.6	+2.5

Hints to Review Questions

1. When two electrons occupy a single atomic orbital, their spins are anti-parallel (or paired). This arrangement is called a singlet state. When two electrons occupy one atomic orbital each, their spins are said to be parallel and this is called a triplet state.
2. Tape recorders, cassettes and floppies are all storage devices. They contain a thin coating of a magnetic substance like iron oxide on a thicker, inert base like plastic. Data is stored on the thin magnetic coating. When placed near a television, the electromagnetic waves emanating from the television interfere with the magnetic field of the coating, thus distorting the data stored on it.

Further Exploration through the Internet

http://www-istp.gsfc.nasa.gov/Education/Imagnet.html
http://www.galaxy.net/~k12/electric/index.shtml
http://www.essex1.com/people/speer/elect.html
http://scitec.uwichill.edu.bb/cmp/online/P10D/p10D.htm
http://micro.magnet.fsu.edu/electromag/java/
http://theory.uwinnipeg.ca/physics/mag/
http://www.wondermagnet.com/dev/magfaq.html

Appendix A

Fundamental SI units

Quantity	*Unit*	*Symbol*	*Definition*
Length	metre	m	The metre is the length of the path travelled by light in vacuum during a time interval of 1/299 792 458 of a second.
Mass	kilogram	kg	The kilogram is the unit of mass; it is equal to the mass of the international prototype of the kilogram.
Time	second	s	The second is the duration of 9 192 631 770 periods of the radiation corresponding to the transition between the two hyperfine levels of the ground state of the caesium 133 atom.
Electric current	ampere	A	The ampere is that constant current which, if maintained in two straight parallel conductors of infinite length, of negligible circular cross-section, and placed 1 metre apart in vacuum, would produce between these conductors a force equal to 2×10^{-7} newton per metre of length.
Thermodynamic temperature	kelvin	K	The kelvin, unit of thermodynamic temperature, is the fraction 1/273.16 of the thermodynamic temperature of the triple point of water.
Amount of substance	mole	mol	The mole is the amount of substance of a system which contains as many elementary entities as there are atoms in 0.012 kilogram of carbon 12. When the mole is used, the elementary entities must be specified. These elementary entities may be atoms, molecules, ions, electrons, other particles, or specified groups of such particles.
Luminous intensity	candela	cd	The candela is the luminous intensity, in a given direction, of a source that emits monochromatic radiation of frequency 540 $\times 10^{12}$ hertz and that has a radiant intensity in that direction of 1/683 watt per steradian.

Source: Bureau International des Poid Mesures, Sèvrés, France (www.bipm.fr).

SI derived units with special names and symbols

Derived Quantity	SI derived unit			
	Name	Symbol	In terms of other SI units	In terms of fundamental SI units
Frequency	hertz	Hz		s^{-1}
Force	newton	N		$m\ kg\ s^{-2}$
Pressure	pascal	Pa	$N\ m^{-2}$	$m^{-1}\ kg\ s^{-2}$
Energy	joule	J	N m	$m^2\ kg\ s^{-2}$
Power	watt	W	$J\ s^{-1}$	$m^2\ kg\ s^{-3}$
Electric charge	coulomb	C		s A
Electric potential	volt	V	$W\ A^{-1}$	$m^2\ kg\ s^{-3}\ A^{-1}$
Capacitance	farad	F	$C\ V^{-1}$	$m^{-2}\ kg^{-1}\ s^4\ A^2$
Electric conductance	siemens	S	$A\ V^{-1}$	$m^{-2}\ kg^{-1}\ s^3\ A^2$
Magnetic flux	weber	Wb	V s	$m^2\ kg\ s^{-2}\ A^{-1}$
Magnetic flux density	tesla	T	$Wb\ m^{-2}$	$kg\ s^{-2}\ A^{-1}$
Inductance	henry	H	$Wb\ A^{-1}$	$m^2\ kg\ s^{-2}\ A^{-2}$
Celsius temperature	degree Celsius	°C		K

SI prefixes

Factor	Prefix	Symbol
10^{-24}	yocto	d
10^{-21}	zepto	c
10^{-18}	femto	m
10^{-15}	micro	μ
10^{-12}	nano	n
10^{-9}	pico	p
10^{-6}	femto	f
10^{-3}	atto	a
10^{-2}	zepto	z
10^{-1}	yocto	y
10^{1}	deca	da
10^{2}	hecto	h
10^{3}	kilo	k
10^{6}	mega	M
10^{9}	giga	G
10^{12}	tera	T
10^{15}	peta	P
10^{18}	exa	E
10^{21}	zetta	Z
10^{24}	yotta	Y

Fundamental Physical Constants

Constant	Symbol	Value in SI units
Speed of light in vacuum	c	2.998×10^{8} m s-1
Elementary charge	e	1.602×10^{-19} C
Avogadro constant	N_A	6.022×10^{23} mol-1
Atomic mass unit	u	1.661×10^{-27} kg
Faraday constant	F	9.649×10^{4} C mol-1
Plank constant	h	6.626×10^{-34} J s
Universal gas constant	R	8.314 J K^{-1} mol^{-1}
Boltzmann constant	k_B	1.381×10^{-23} J K^{-1}
Gravitational constant	G	6.673×10^{-11} N m^{2} kg^{-2}
Electron rest mass	m_e	9.110×10^{-31} kg
Proton rest mass	m_p	1.673×10^{-27} kg
Permeability of vacuum	μ_0	12.566×10^{-7} T^{2} m^{3} J^{-1}
Permittivity of vacuum	ε_0	8.854×10^{-12} C^{2} J^{-1} m^{-1}
Bohr magneton	μ_B	9.274×10^{-24} J T^{-1}
Nuclear magneton	μ_N	5.051×10^{-27} J T^{-1}

Conversion factors

CGS units	SI units
1 cal	4.184 J
1 erg	10^{-7} J
1 inch	2.54 cm
1 pound	0.454 kg
1 yard	0.9144 m
1 angstrom	10^{-10} m
1 atm = 1.013 bar = 760 torr	1.013×10^{5} N m^{-2}
1 mile	1.6 km
1 dyne	10^{-5} N
1 gauss	10^{-4} T
1 poise	0.1 Pa s

The Greek alphabet

English Name	Case	
	Upper	Lower
Alpha	Α	α
Beta	Β	β
Gamma	Γ	γ
Delta	Δ	δ
Epsilon	Ε	ε
Zeta	Ζ	ζ
Eta	Η	η
Theta	Θ	θ
Iota	Ι	ι
Kappa	Κ	χ
Lambda	Λ	λ
Mu	Μ	μ
Nu	Ν	ν
Xi	Ξ	ξ
Omicron	Ο	ο
Pi	Π	π
Rho	Ρ	ρ
Sigma	Σ	σ, *S*
Tau	Τ	τ
Upsilon	Υ	υ
Phi	Φ	φ
Chi	Χ	χ
Psi	Ψ	ψ
Omega	Ω	ω

Sigma (□, □): There are two forms for the letter Sigma. When written at the end of a word, it is written like this: □. If it occurs anywhere else, it is written like this: □.

Upsilon (□): We suggest that you pronounce this letter like "u" in "put". The preferred pronunciation is actually more like the German "ü" as in "Büchner", or like the French "u" as in "tu".

Chi (□): This sounds like "ch" in "Bach", and not like "ch" in "chair".

International Atomic Weights

A number in parentheses indicates the uncertainty in the last digit of the atomic weight.

The value enclosed in brackets, e.g. [209], indicates the mass number of the longest-lived isotope of the element. The element has no stable nuclides.

At No	Symbol	Name	Atomic Wt
89	Ac	Actinium	[227]
13	Al	Aluminium	26.981538(2)
95	Am	Americium	[243]
51	Sb	Antimony	121.760(1)
18	Ar	Argon	39.948(1)
33	As	Arsenic	74.92160(2)
85	At	Astatine	[210]
56	Ba	Barium	137.327(7)
97	Bk	Berkelium	[247]
4	Be	Beryllium	9.012182(3)
83	Bi	Bismuth	208.98038(2)
107	Bh	Bohrium	[264]
5	B	Boron	10.811(7)
35	Br	Bromine	79.904(1)
48	Cd	Cadmium	112.411(8)
55	Cs	Caesium	132.90545(2)
20	Ca	Calcium	40.078(4)
98	Cf	Californium	[251]
6	C	Carbon	12.0107(8)
58	Ce	Cerium	140.116(1)
17	Cl	Chlorine	35.4527(9)
24	Cr	Chromium	51.9961(6)
27	Co	Cobalt	58.933200(9)
29	Cu	Copper	63.546(3)
96	Cm	Curium	[247]
105	Db	Dubnium	[262]
66	Dy	Dysprosium	162.50(3)
99	Es	Einsteinium	[252]
68	Er	Erbium	167.26(3)
63	Eu	Europium	151.964(1)
100	Fm	Fermium	[257]
9	F	Fluorine	18.9984032(5)
87	Fr	Francium	[223]
64	Gd	Gadolinium	157.25(3)

31	Ga	Gallium	69.723(1)
32	Ge	Germanium	72.61(2)
79	Au	Gold	196.96655(2)
72	Hf	Hafnium	178.49(2)
108	Hs	Hassium	[265]
2	He	Helium	4.002602(2)
67	Ho	Holmium	164.93032(2)
1	H	Hydrogen	1.00794(7)
49	In	Indium	114.818(3)
53	I	Iodine	126.90447(3)
77	Ir	Iridium	192.217(3)
26	Fe	Iron	55.845(2)
36	Kr	Krypton	83.80(1)
57	La	Lanthanum	138.9055(2)
103	Lr	Lawrencium	[262]
82	Pb	Lead	207.2(1)
3	Li	Lithium	6.941(2)
71	Lu	Lutetium	174.967(1)
12	Mg	Magnesium	24.3050(6)
25	Mn	Manganese	54.938049(9)
109	Mt	Meitnerium	[268]
101	Md	Mendelevium	[258]
80	Hg	Mercury	200.59(2)
42	Mo	Molybdenum	95.94(1)
60	Nd	Neodymium	144.24(3)
10	Ne	Neon	20.1797(6)
93	Np	Neptunium	[237]
28	Ni	Nickel	58.6934(2)
41	Nb	Niobium	92.90638(2)
7	N	Nitrogen	14.00674(7)
102	No	Nobelium	[259]
76	Os	Osmium	190.23(3)
8	O	Oxygen	15.9994(3)
46	Pd	Palladium	106.42(1)
15	P	Phosphorus	30.973761(2)
78	Pt	Platinum	195.078(2)
94	Pu	Plutonium	[244]
84	Po	Polonium	[209]
19	K	Potassium	39.0983(1)

59	Pr	Praseodymium	140.90765(2)
61	Pm	Promethium	[145]
91	Pa	Protactinium	231.03588(2)
88	Ra	Radium	[226]
86	Rn	Radon	[222]
75	Re	Rhenium	186.207(1)
45	Rh	Rhodium	102.90550(2)
37	Rb	Rubidium	85.4678(3)
44	Ru	Ruthenium	101.07(2)
104	Rf	Rutherfordium	[261]
62	Sm	Samarium	150.36(3)
21	Sc	Scandium	44.955910(8)
106	Sg	Seaborgium	[263]
34	Se	Selenium	78.96(3)
14	Si	Silicon	28.0855(3)
47	Ag	Silver	107.8682(2)
11	Na	Sodium	22.989770(2)
38	Sr	Strontium	87.62(1)
16	S	Sulfur	32.066(6)
73	Ta	Tantalum	180.9479(1)
43	Tc	Technetium	[98]
52	Te	Tellurium	127.60(3)
65	Tb	Terbium	158.92534(2)
81	Tl	Thallium	204.3833(2)
90	Th	Thorium	232.0381(1)
69	Tm	Thulium	168.93421(2)
50	Sn	Tin	118.710(7)
22	Ti	Titanium	47.867(1)
74	W	Tungsten	183.84(1)
92	U	Uranium	238.0289(1)
23	V	Vanadium	50.9415(1)
112	Uub	Ununbium	[277]
116	Uuh	Ununhexium	[289]
110	Uun	Ununnilium	[269]
118	Uuo	Ununoctium	[293]
114	Uuq	Ununquadium	[285]
111	Uuu	Unununium	[272]
54	Xe	Xenon	131.29(2)
70	Yb	Ytterbium	173.04(3)
39	Y	Yttrium	88.90585(2)
30	Zn	Zinc	65.39(2)
40	Zr	Zirconium	91.224(2)

Appendix B

The Periodic Table

Period	1 IA 1A	2 IIA 2A	3 IIB 3B	4 IVB 4B	5 VB 5B	6 VIB 6B	7 VIIB 7B	8	9 VIII 8	10	11 IB IB	12 IIB 2B	13 IIIA 3A	14 IVA 4A	15 VA 5A	16 VIA 6A	17 VIIA 7A	18 vIIA 8A
1.	1 H 1.008																	2 He 4.002
2.	3 Li 6.941	4 Be 9.012											5 B 10.81	6 C 12.011	7 N 14.006	8 O 15.999	9 F 18.998	10 Ne 20.179
3.	11 Na 22.989	12 Mg 24.305											13 Al 26.981	14 SI 28.085	15 P 30.973	16 S 32.066	17 Cl 35.452	18 Ar 39.948
4	19 K 39.098	20 Ca 40.078	21 Sc 44.956	22 Ti 47.88	23 V 50.941	24 Cr 51.996	25 Mn 54.938	26 Fe 55.847	27 Co 58.933	28 Ni 58.69	29 Cu 63.546	30 Zn 65.39	31 Ga 69.723	32 Ge 72.61	33 As 74.921	34 Se 78.96	35 Br 98.904	36 Kr 39.948
5.	37 Rb 85.467	38 Sr 87.62	39 Y 88.906	40 Zr 91.224	41 Nb 92.906	42 Mo 95.94	43 Tc 98.906	44 Ru 101.07	45 Rh 102.905	46 Pd 106.42	47 Ag 107.868	48 Cd 112.411	49 In 114.82	50 Sn 118.71	51 Sb 121.75	52 Te 127.6	53 I 126.904	54 Xe 83.8
6.	55 Cs 132.905	56 Ba 137.327	57* La 138.905	72 Hf 178.49	73 Ta 180.947	74 W 183.85	75 Re 186.207	76 Of 190.2	77 Ir 192.22	78 Pt 195.08	79 Au 196.966	80 Hg 200.59	81 Ti 204.383	82 Pb 207.2	83 Bl 208.98	84 Po (209.98)	85 At (210)	86 Rn 131.29
7.	87 Fr 223.02	88 Ra 226.025	89** Ac 227.03	104 Rf (261.11)	105 Db (262.114)	106 Sg (263)	107 Bh (262)	108 Hs (265)	109 Mt (266)	110 Uun (272)	111 Uuu (272)	112 Uub (227)	113 Uut (0)	114 Uuq (285)	115 Uup (0)	116 Uuh (089)	117 Uus (0)	118 Uuo (1293)

*Lanthanide series 4f	58 Ce 140.115	59 Pr 140.907	60 Nd 144.24	61 Pm 145	62 Sm 150.36	63 Eu 151.965	64 Gd 157.25	65 Tb 158.925	66 Dy 162.5	67 Ho 164.93	68 Er 167.26	69 Tm 168.934	70 Yb 173.04	71 Lu 174.967
**Actinide Series 5f	90 Th 232.038	91 Pa 231.035	92 U 238.028	93 Np 237.048	94 Pu (244.06)	95 Am (243.06)	96 Cm (247.07)	97 Bk (247.07)	98 Cf (251.08)	99 Es (254.08)	100 Fm (257)	101 Md 258	102 No (259.101)	103 Lr (260.105)

Atomic number	80
Atomic Symbol	**Hg**
Relative mass	200.5

The names of the elements 110-118 are the Latin equivalents of these numbers.

Appendix C

M Sc I Practical Syllabus

I. Thermodynamics and the Phase Rule

1. Determination of partial molal volume of methanol in dilute aqueous solutions.
2. Determination of heat of solution of benzoic acid/salicylic acid by solubility measurements.
3. Study of three-component system: water-acetic acid-chloroform/water-toluene-acetic acid.
4. Study the variation of the solubility of calcium sulphate with ionic strength and hence determine the thermodynamic solubility product (complexometric titration with EDTA).
5. Determination of the equilibrium constant of the reaction $KI + I_2 \longleftrightarrow KI_3$.

II. Reaction Kinetics

1. Investigation of the reaction between acetone and iodine.

III. Conductometry

1. Titration of a mixture of trichloroacetic acid, monochloroacetic acid and acetic acid by titrating with sodium hydroxide conductometrically.
2. Verification of Ostwald's dilution law and determination of the dissociation constant of a weak monobasic acid conductometrically.
3. Determination of the concentrations/amounts of sulphuric acid, acetic acid and copper sulphate by conductometric titration with sodium hydroxide.
4. Determination of the concentration/amount of oxalic acid by conductometric titration with sodium hydroxide.

IV. Potentiometry

1. Determination of the concentrations/amounts of iodide, bromide and chloride in the mixture by potentiometric titration with silver nitrate.
2. Determination of the solubility product of silver chloride potentiometrically using a chemical cell.
3. Determination of the solubility product of silver chloride potentiometrically using a concentration cell.

4. Determination of pK values of maleic acid/malonic acid by potentiometric titration with sodium hydroxide using glass electrode.
5. Determination of pK values of phosphoric acid by potentiometric titration with sodium hydroxide using glass electrode.

V. Colorimetry

1. Determination of the composition and stability constant of a complex formed between iron (III) ions and salicylic acid by Job's method of continuous variation.

M Sc II Practical Syllabus

Group A

Distribution Methods

1. Determination of the formula of the complex formed between silver and ammonia.
2. Determination of the formula of the complex formed between copper (II) ion and ammonia.

Solubility

1. Study the variation in solubility of calcium hydroxide in the presence of sodium hydroxide, and hence determine the solubility product of calcium hydroxide at room temperature.
2. Study the variation in the solubility of potassium hydrogen tartrate with ionic strength and hence determine the mean ionic activity coefficient.

Phase Equilibrium

1. Construct a phase diagram of a simple eutectic system of naphthalene and biphenyl.
2. Construct a phase diagram of a simple eutectic system of *o*-nitrophenol and *p*-toluidine.
3. Determine the transition temperature of a hydrated salt.

Group B

Reaction Kinetics

1. Study the kinetics of hydrolysis of *t*-butyl chloride by conductometric measurements.

2. To determine the energy of activation of hydrolysis of methyl acetate catalysed by hydrochloric acid.
3. To study the influence of ionic strength on reaction between potassium persulphate and potassium iodide in solution.
4. Study the autocatalytic reaction between potassium permanganate and oxalic acid.
5. Investigate the reaction between potassium persulphate and potassium iodide by colorimetric measurements.

Cryoscopy

1. Determine the cryoscopic constant for water using a substance of known relative molecular mass and determine the relative molecular mass of a non-electrolyte.
2. Determine the mean ionic activity coefficient of HCl by the equilibrium freezing point depression method.
3. Determine the mean ionic activity coefficient of KCl and $BaCl_2$.

Macromolecules

1. Determine the limiting viscosity number of polystyrene.
2. Determine the chain linkage in poly(vinyl alcohol) from viscosity measurements.
3. Determine the relative molecular mass of polystyrene from viscometry measurements.
4. Determine the relative molecular mass of hydroxyl terminated polybutadiene and carboxyl terminated polybutadiene.
5. Determination of radius of a molecule by viscosity measurements.

Group C

Surface chemistry and colloids

1. Determine the Critical Micelle Concentration of sodium lauryl sulphate/ N-cetyl-N,N,N-trimethyl ammonium bromide (CTAB) from measurements of surface tension and conductivities at different concentrations.
2. Determination of flocculation values for negative, positive, suspendoid sols using electrolytes having cations and anions of different valencies in order to test the validity of the Hardy-Schulze Rule.

Transport Number

1. Determine the transport number of copper ions in copper sulphate by the Hittorf method.

2. Determine the transport number of silver ions in silver nitrate by the Hittorf method.
3. Determine the transport number of hydrogen ions in hydrochloric acid by the Hittorf / moving boundary method.

Potentiometry/pH-metry

1. Determine the formula of silver-ammonia complex ions.
2. Determine the stability constant of the silver-ammonia complex ions.
3. Determine the transport number of silver and nitrate ions in aqueous solution from the cell potential of the concentration cell with junction potential.
4. Determination of Hammett constant of *o*-, *m*-, *p*-amino/nitro benzoic acid by pH measurements.
5. Determine the acidic and basic dissociation constants of an amino acid and hence the isoelectric point of the acid.
6. Determination of the proton-ligand stability constant of an organic acid and the metal-ligand stability constant of its complex by pH measurements.

Voltammetry

1. Determine the half-wave potential of Cd(II), Cu(II) and Zn(II) ions in 0.1M KCl solution.
2. Study the complex formation between
   - Pb (II) ions and malonic acid
   - Cd (II) ions and glutamic acid.
3. Determine the concentration of Pb(II) in solution by amperometric titration with potassium dichromate solution.
4. Study the redox behaviour of $K_3[Fe(CN)_6]/K_4[Fe(CN)_6]$ by cyclic voltammetry.

Spectrophotometry

1. Study the complex formation between Ni ions and 1,10-phenanthroline.
2. Determine the ionisation constant of methyl red/bromophenol blue and determine the pH value of the given buffer solutions.
3. Determine the stability constant of the complex formed between Fe(III) and 5-sulphosalicylic acid at pH 2 and pH 3 by colorimetric measurements.
4. Determine the stability constant of the complex formed between Cu(II) ions and 5-sulphosalicylic acid at various pH by spectrophotometric

method and study the complex formation between pH 3-5 and 7-9 and calculate the equilibrium constant at these pH values and hence find free energy of formation.

Magnetic Susceptibility

1. Determination of the magnetic susceptibility of iron(II) ammonium sulphate (Mohr's salt) or $K_3[Fe(CN)_6]$ by Guoy's method / Faraday method.

Group D

Interpretation of Data

1. Analysis of rotational-vibrational spectra of rigid and non-rigid diatomic molecules.
2. Analysis of electronic spectra of diatomic molecules.
3. Analysis of vibrational-electronic spectra of simple polyatomic molecules.
4. Analysis of NMR, ESR spectra.
5. Analysis of X-Ray powder diffraction pattern (cubic system).
6. Analysis of DTA, TG, DSC curves.
7. M. O. calculations for simple molecules using computer program.

Index